SOME BIOLOGICAL
TECHNIQUES IN
ELECTRON MICROSCOPY

CONTRIBUTORS

WILLIAM G. BANFIELD
STANLEY BULLIVANT
W. W. HARRIS
D. F. PARSONS
JELLE C. RIEMERSMA

SOME BIOLOGICAL TECHNIQUES IN ELECTRON MICROSCOPY

Edited By D. F. PARSONS
Roswell Park Memorial Institute
Buffalo, New York

 1970

ACADEMIC PRESS   New York and London

ACADEMIC PRESS, INC.
111 Fifth Avenue, New York, New York 10003

United Kingdom Edition published by
ACADEMIC PRESS, INC. (LONDON) LTD.
Berkeley Square House, London W1X 6BA

LIBRARY OF CONGRESS CATALOG CARD NUMBER: 72-107549

PRINTED IN THE UNITED STATES OF AMERICA

284778

CONTENTS

LIST OF CONTRIBUTORS

Numbers in parentheses indicate the pages on which the authors' contributions begin.

WILLIAM G. BANFIELD,* Laboratory of Pathology, National Cancer Institutes of Health, Bethesda, Maryland (166)

STANLEY BULLIVANT, Cell Biology Department, University of Auckland, Auckland, New Zealand (101)

W. W. HARRIS, Molecular Anatomy (MAN) Program, Oak Ridge National Laboratory, Oak Ridge, Tennessee (147)

D. F. PARSONS, Roswell Park Memorial Institute, Buffalo, New York (1)

JELLE C. RIEMERSMA, Laboratory of Medical Chemistry, University of Leiden, Leiden, Netherlands (69)

* Present address: Fission Products Inhalation Lab, Sandia Base, Albuquerque, New Mexico.

PREFACE

This volume was initiated to focus attention on some of the problems that prevent or hinder progress in biological electron microscopy. The best resolution of available commercial microscopes (4 Å point-to-point or better) far exceeds the resolution that can be obtained on an actual biological specimen. This waste of the instrument potential represents a serious loss in biological research capability, and is of concern to many biologists, biophysicists, and medical researchers.

The electron microscope has great potential for solving many pressing medical and biological research problems. Only instruments which utilize imaging to obtain their analytical output can give large quantities of information per experiment. A high rate of output-information transfer is necessary for analyzing such complicated structures as cell membranes and chromosomes in a reasonable number of man-years. The electron microscope uniquely meets these requirements. The image is obtained directly without the necessity for computing it as in x-ray crystallography. Direct imaging has the additional advantage that crystals are not required and single particles and amorphous aggregates can be visualized just as well.

Because of these attractive features as a biological research tool, it is very disappointing that current electron microscope research has come up against some severe obstacles. Adequate visualization by scanning or transmission microscopy of single molecules has still not been achieved nor the resolution of single atoms. The problem is mainly one of contrast. To progress toward the single atom level of resolution the old techniques of introducing heavy metal atoms to increase contrast must be abandoned. All such techniques lead to the irregular deposition of relatively large crystallites without enhancing the contrast of individual atoms. We must now turn to electron optical methods of obtaining contrast such as dark field or phase contrast. This is discussed in the chapter Problems in High Resolution Electron Microscopy of Biological Materials in Their Natural State.

Even when a satisfactory electron optical method of obtaining contrast is available, major problems remain. The frequent absence of an electron diffraction pattern from ordered biological materials shows that the combined effects of drying out in the vacuum and damage due to the electron beam have seriously impaired the high resolution imaging possibilities. At the moment, it appears that decreased beam damage will require higher acceleration voltages and use of image intensifiers. The distortion produced by drying requires either special specimen hydration chambers or improved methods of handling frozen materials.

It has always appeared to electron microscopists that freezing the specimen should be a good way of preparing it for the electron microscope. However, unexpected difficulties have been encountered. The present state of the art is given in the chapter by S. Bullivant.

Recently, with the improved modern microscopes available, it has been realized that in many cases chromatic aberration due to energy losses in the specimen is a more important limitation of resolution than spherical aberration. For high resolution studies on thin objects a very thin support film is required. In addition, it must be possible to distinguish the structure of the object from that of the support film. This subject is surveyed in the chapter by W. W. Harris.

In cytological studies in which the classic thin section method is still giving new results, there has been an increase in our understanding of the effects of chemical fixation (see the chapter by J. C. Riemersma). Also, a start has been made toward automating the thin-section processing technique (chapter by W. G. Banfield).

I particularly thank all the contributors for their efforts and their patience. I thank Academic Press for their understanding and cooperation in publishing this work.

SOME BIOLOGICAL
TECHNIQUES IN
ELECTRON MICROSCOPY

Chapter I

PROBLEMS IN HIGH RESOLUTION ELECTRON MICROSCOPY OF BIOLOGICAL MATERIALS IN THEIR NATURAL STATE

D. F. PARSONS

I. Introduction

A major goal of biological electron microscopy is the visualization of atoms in biological molecules and structures. This goal requires a resolution of about 1 Å, adequate contrast, and a method of keeping the specimen in an intact state during examination in the electron microscope. It is also required to rigidly support the specimen in such a way that the support film structure does not interfere with the specimen structure. Thus, the goal of atomic resolution of intact biological structures presents some very challenging physical problems, more problems, in fact, than does high resolution electron microscopy of metals and other stable inorganic substances. On the other hand, the solution of these problems might lead to valuable observations in biology and medicine. These could shortcut some of the difficult biochemical work required to establish normal cell structure and the differences between normal and pathological cell structures.

The development of techniques for preparing biological specimens for electron microscopy has, in the past, lagged behind the improvements in resolution of the commercially available electron microscopes. Recent models of electron microscopes can be induced (at least occasionally!) to give point-to-point resolutions of 3 Å on certain kinds of tangentially metal-shadowed support films. However, resolution tests on stable, high contrast metal-shadowed specimens are very different from high resolution experiments of unstable, low contrast, biological specimens. Apart from the delicate nature of biological specimens compared to inorganic specimens and the need to minimize damage due to vacuum drying and the electron beam, there is the need to enhance the contrast of the image. Biological electron microscopy has become more dependent on finding new techniques for enhancing contrast than on improving microscope resolution. Most effort has been expended in enhancing the scattering power of particular regions of the specimen by the introduction of heavy metal atoms rather than in obtaining the contrast electron optically. All methods of metal contrasting presently available limit the resolution to not more than 10 Å, and involve either vacuum drying or chemical damage of the specimen. Positive staining requires chemical reaction with heavy metal salts with an uncertain degree of damage to the specimen. Course (10 Å or more) deposits of heavy metal atoms are formed in the structure. Only the metal deposits are visible and the rest of the structure is very low in contrast and almost invisible. The positive staining technique has very limited potential as far as the aim of atomic resolution is concerned. The situation is no better for negative staining, since this technique is also limited by stain granularity to about 10 Å resolution and the inner hydrophobic or nonpenetrated portions of the structure are not shown. Metal shadowing is limited by a granularity or crystallinity of the deposited metal of 10–12 Å dimension, and by the fact that only surfaces of structures are seen. At this point in the development of biological electron microscopy, it appears likely that microscopes will soon routinely reach the 1 Å resolution level, but something quite different must be done about both specimen contrast and about specimen damage. In this report we shall shift the emphasis away from specimen preparation toward modification of the electron microscope instrument to produce more contrast electron optically and to reduce specimen damage during their examination.

On the assumption that the specimen contrast and specimen damage problems can be overcome, the limitations on microscope resolution will be discussed in some detail. Our understanding of electron imaging is still very rudimentary. Many concepts have been adopted from light microscope theory without due regard to their applicability. A quantitative reexamination of the electron scattering effects observed with organic and biological materials and of image formation is long overdue. In discussing this question it becomes clear that an exact description of the aberrations produced by magnetic electron lenses is essential. The crudity of the present approach can be recognized by the attempts to describe aberrations of the very defective magnetic lens by a single term expression, whereas three terms, or more, are required to describe the aberrations of the more perfect light microscope objectives. In addition, the advent of the information theory approach to lens action *(93)* now makes it possible to describe the total wave aberration of the magnetic lens by several different kinds of optical transfer functions. There is no longer any need to break the aberrations down into spherical, astigmatic, or distortion components. Quantitative tests of electron imaging theories, using simple objects, are very rare in the literature. The explanation probably lies in the ultrarapid development of practical electron microscopy in which empirical approaches were sufficient. However, now that such approaches have ceased to provide easy solutions to problems of biological electron imaging, it is time to reassess the quantitative status of the art of electron imaging and to attempt to put in on a sounder theoretical basis.

II. Resolution Limitations of Present Day
Electron Microscopes

The rate of development of the electron microscope was influenced, to some extent, by mistaken concepts about the resolution limit of available instruments. Up to about 1945, for example, it was assumed that spherical aberration was the practical resolution limitation, whereas Hillier and Ramberg (67) showed that, at this time, astigmatism was a far more important limitation. By compensating the astigmatism by means of magnetic shims, they immediately improved the resolution from 20–25 Å to 10 Å. Now that astigmatism can be accurately and quickly compensated, it is again commonly believed that spherical aberration is the factor that limits the practical resolution to 2–3 Å. In this section we shall discuss other types of aberration that may, in fact, limit the resolution.

It was natural to apply the Abbe concepts of image formation to the electron microscope. However, as pointed out by Glaser (49), this was often done without recognizing the essential differences that exist between light and electron microscopes. In the Abbe approach the resolution is related to the highest order reflection, or highest spatial frequency, appearing in the back focal plane of the lens. Since light can be diffracted 90° or more by periodicities in the object, the resolution of a light microscope objective is limited by its ability to collect and pass the higher order diffraction spectra. However, such a concept of a limiting aperture for the lens is not applicable to electron lenses. Because of the much shorter wavelength, diffraction of electrons does not occur outside of an angle about 5° for conventional (40–100 kV) electron microscope voltages. Since the object in an electron microscope is either placed some distance in front of the front principal plane, or between the (crossed) front and back principal planes, a 5° cone of diffracted rays will be passed by any practical magnetic lens. It is observed that both low and high resolution microscopes give back focal plane diffraction patterns of single crystals (selected area diffraction) which show spacings to about 0.6 Å. The pattern is not cut off by the lens at this spatial frequency, but by the temperature factor (or thermal vibration of atoms). In addition, it is readily observed, by comparing back focal plane diffraction patterns (i.e., produced by the objective lens) with diffraction patterns produced without imaging lenses, that the intensity distribution of the wide angle reflections is not diminished by passage through the lens. Since a resolution of 0.6 Å has not yet been achieved in an electron microscope, it is clear that the Abbe concept (that resolution is limited by the numerical aperature of the objective) does not apply to magnetic lenses without an objective aperture. A comparison between a lens-produced diffraction pattern (gold film) and one produced without imaging lenses is shown in Figs. 1A and 1B. The lens-produced patterns show no loss in intensity, but they are slightly less distinct. Quantitatively, the slight broadening and diffuseness observed in the lens-produced pattern cannot be ascribed to spherical aberration of the objective lens, but may be due to sperical aberration and astigmatism of the long focal length intermediate lens used. (The Siemens Elmiskop Ia has no stigmator for the intermediate lens.) For lenses having larger aberrations than magnetic lenses (e.g., electrostatic lenses) there is a measurable and increasing displacement of the reflections toward the axis due to spherical aberration (*52, 68, 88*) for reflections of increasing inclination to the lens axis.

A. Dependence of Imaging Theory on a Detailed Description of Lens Aberrations

Available experimental methods for determining of the aberrations of magnetic lenses are not entirely satisfactory. Most techniques [e.g., the Hartmann diaphragm method, the diffraction diagram method, straight edge method, etc. (*52*)] are only applicable to electrostatic lenses that have spherical aberration constants some 20 times greater than magnetic lenses.

In discussions about spherical aberration of magnetic lenses, a single term expression has usually been used. A ray making an angle α with the axis at the object plane arrives at the Gaussian image plane with a displacement from the axis of Δr, given by:

$$\Delta r = MC_o\alpha^3 \qquad (1)$$

where M is the magnification, C_o is the spherical aberration coefficient

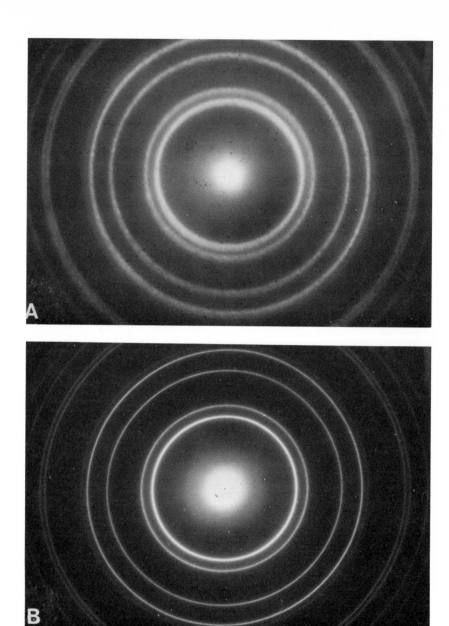

Fig. 1. Comparison of the electron diffraction pattern of a gold film obtained with and without the use of an objective lens. (A) Obtained using a Siemens Elmiskop Ia electron microscope with an objective current of 680 mA (80 kV).

of the objective, and α is the scatter angle or deviation from the optical axis. With reference to the object plane, each object point appears to be smeared out to a disc of radius:

$$\Delta r' = C_o \alpha^3 \tag{2}$$

The disc of confusion can be reduced in size by slight defocusing to find the minimum diameter of the envelope or caustic of focused rays. This gives approximately:

$$\Delta r'' = 1/4 C_o \alpha^3 \tag{3}$$

It cannot be supposed that such expressions accurately describe the spherical aberration for all values of α, since the aberration is very large for large α (hundreds of wavelengths for maximum α). A more likely expression would be:

$$\Delta r'' = 1/4(C_o \alpha^3 + C'_o \alpha^5 + C''_o \alpha^7, \text{etc.}) \tag{4}$$

In any case this approach emphasizes a geometrical description of a single aberration. However, it would be more useful to use the information theory and physical optics approach (*93*), which gives the total wave aberration for each frequency component or spatial frequency in the Fourier transform of the three-dimensional distribution of scattering points of the object.

The manner in which the electron microscope manufacturer has arrived at the value for the spherical aberration constant, C_o, for their objective lens is usually not clearly described. In most cases it is likely that it has been calculated from the axial magnetic field distribution, although some more thorough experimental determinations have been carried out by Ruska (*104*), Becker and Wallraff (*5*), and Kunath and Riecke (*80*). Calculations have been made for various standard field distributions of magnetic lenses (*101*). These are subject to the errors involved in the measurement of the axial field dis-

---

The near back focal plane diffraction pattern was photographed in the selected area diffraction mode (No. 3388). (B) Objective, intermediate, and projector lenses off and wide bore projector pole piece in place (simple diffraction camera arrangement without imaging lenses) (No. 3392). The diffraction pattern with imaging lenses on could not be made as sharp as with the lenses off due to lens aberrations. However, the reflections in both cases (with and without objective lens) go out to at least 0.7 Å indicating that resolution is not limited by the numerical aperture of the objective lens.

tribution and the approximations involved in using paraxial ray tracing formulas. For example, the aberration has been calculated for symmetrical bell-shaped fields (48) and asymmetrical bell-shaped fields (26). The dependence of the spherical aberration coefficient on object position and magnification has recently been emphasized by Hawkes (58).

Kunath and Riecke (80) have experimentally measured the transverse spherical aberration of the Siemens Elmiskop using a different approach. The displacement of the image on tilting the beam is measured. The tilt causes an originally axial image to be formed by rays that pass through the periphery of the lens. The authors estimated the second term in an expanded expression for spherical aberration and concluded that it was small.

It appears that a measurement of the axial field distribution (H_z) should serve to describe the whole magnetic field of the lens, provided the Laplacian expression applies. Similarly, such a measurement should enable exact calculation of electron trajectories and lens aberrations provided it can be made accurately enough. Measurements have been made with vibrating probes (32), magnetoresistive effect in small bismuth wires (81) using coils of OD 0.75 mm (75), and also very small coils [100 turns of 0.04 cm diameter and 0.05 cm in length (23)]. The axial value of magnetic field intensity, H_z, and its first and second differentials, H'_z and H''_z, may be used to calculate the spherical aberration coefficient C_o (50):

$$C_o = \frac{e}{96m_0V} \int_{z_0}^{z_1} \left[\frac{2e}{m_0V} H_z{}^4 + 5H'_z{}^2 - H_zH''_z \right] y^4 \, dz \qquad (5)$$

where m_0 = rest mass of electron, e = electron charge, z_0 = center of lens, z_1 = edge of lens where $H_z = 0$, V = accelerating voltage, and $y(z)$ = displacement of electron trajectory from the axis at a distance z from lens center. Results of this calculation for different lens geometries are given by Dugas et al. (29) [see also Durandeau and Fert (35)].

The accuracy of such calculations depends first on the errors in measuring the axial field distribution. These may be large since magnetic probes of sufficiently small volume have not been used. Critical studies of the possible errors involved in calculations of the type of Eq. (5) need to be carried out. An improved miniature magnetic field strength probe of micron dimensions [e.g., a micro-Hall effect probe (*109*)] opens up interesting possibilities for accurate computer calculation of aberrations, magnetic lens imaging, and experiments on electron imaging. The effect of the wave aberrations (including time dependent phase shifts due to power supply instabilities and AC ripple) can be studied with respect to image contrast and also resolution.

For imaging calculations, the most useful presentation of aberration data would be the phase shift of the peripheral ray with respect to the axial ray for different displacements, y from the axis, since the image contrast and resolution depend critically on this. There is no doubt that if more accurate magnetic field distribution data were available, this could be computed without making any of the significant approximations usually made in paraxial ray tracing expressions. It is also possible that detailed experimental ray tracing can be carried out. In our laboratory, H. M. Johnson and D. F. Parsons made use of the images of the beam produced by controlled beam contamination to examine the profile and intensity distribution of the beam at points close to the back focal plane. Japanese workers have recently made use of the fine granularity and broad range of intensity response of photoresist for the same purpose (*74*).

B. DYNAMIC AND STATIC LENS ABERRATIONS

Both theoretical and experimental ray tracing experiments that give the wave aberration phase shifts as a function of distance from the axis do not include the dynamic or time dependent phase shifts present in the imaged beam.

The time dependent wave aberration can be thought of as made up of two parts: (1) a random instability in focal length due to random supply fluctuations in the high voltage and the objective lens currents; and (2) a sinusoidal fluctuation due to AC ripple in the high voltage supply and lens current. Electron microscope manufacturers have paid surprisingly little attention to the effects of AC ripple and its elimination. Part of the problem arises from a con-

fusion of terms used in the definition of the stability of a power supply
The chromatic aberration coefficient, C_c, is given by:

$$C_c = \frac{\Delta r}{\alpha} \cdot \frac{V}{\Delta V} \qquad (6)$$

where Δr is the radius of the disc of confusion (referred to the specimen plane) for an energy spread ΔV in a beam of mean voltage V,
and scattered at an angle α to the lens axis. Manufacturers are not
usually specific about whether the quoted high voltage stability
($\Delta E/E$) includes 60 cps ripple or not. Often the chief concern seems
to be stability from drift in an average level of voltage over a period
of some hours rather than the short term fluctuations, including 60
cps ripple, occurring during an exposure time of five seconds. A
quantitative theoretical and experimental investigation of the effect
of 60 cps ripple on image contrast and resolution has yet to be made.
However, there are already many indications that it is important to reduce the AC ripple levels below present levels. To do this effectively would require a completely different arrangement of the electronic supplies and of the wiring. In present microscope designs there are
many sources of ripple, both in the high voltage and the lens current
supplies. The gun filament may be heated by AC. The high voltage
regulator may have AC heated regulator tubes. The reference batteries
or reference voltage supplies are inadequately shielded from AC fields.
In both tube and solid state supply circuits the equipment is often
crowded together, and the regulated DC output lines placed adjacent to
AC lines. Regardless of the manufacturers efforts, the lens supplies, by
the time they reach the console of the microscope, may have picked up
a great deal of AC ripple. Finally, there is a real question as to
whether it is possible for electronic supplies to give sufficiently low
levels of AC ripple and whether wet cell batteries would be better.
A microscope in which particular attention is paid to AC ripple would
require a design where all AC supplies and fields are separated from
DC supplies. Presumably, simplification could be achieved by operating all of the microscope components (e.g., mechanical pump, diffusion
pumps, relays, and panel lights) off a 110 V DC supply rather than
AC.

The effect of voltage and objective supply AC ripple can be regarded as a superimposition of (two) (unrelated) fluctuations in focus.

The focus change associated with the voltage fluctuation ΔV, Eq. (6), is approximately:

$$\Delta f' = -f \cdot \frac{\Delta V}{V} \tag{7}$$

and the focus change due to an unrelated objective current fluctuation ΔI in I A is:

$$\Delta f'' = -2f \cdot \frac{\Delta I}{I} \tag{8}$$

The total focus fluctuation is:

$$\Delta f = -f \left(\frac{\Delta V}{V} + 2 \frac{\Delta I}{I} \right) \tag{9}$$

Since the effects of fluctuations in accelerating voltage and objective current can be made opposite in effect, a possibility exists for modifying the ripple levels so that one balances the other. Hence, one possibility for neutralizing ripple levels would be to monitor, by sensitive C.R.O. display, the hum on all supplies, and to employ phase and amplitude adjusting circuits to keep the ripple at negligible levels.

The effect of ripple on the aberration of the higher spatial frequencies transmitted by the objective lens can be studied theoretically using the information theory approach.

In the information theory approach to imaging, the lens is regarded as modifying and limiting the frequency distribution of the Fourier transform of the object, $F(\omega)$, where ω is the reciprocal periodicity, $1/d$, or *spatial frequency*. The operator which modulates the wave passing through the lens is called the *optical transfer function* or *frequency response function* $H(\omega)$. The operator has two effects: It impresses the lens aberration on the wave; and it weighs the amplitude and phase of the spatial frequencies being transmitted. In light lenses, with small aberrations, the operator $H(\omega)$ is linear with respect to spatial frequency, and the lens can be regarded as a linear filter of spatial frequencies. This is not the case with electron microscope lenses because in the wide angle region the aberrations are much greater than for light lenses. The wave aberration (deviation in wavelengths from a spherical reference sphere) is less than $\lambda/4$ for a high quality light objective resolving about one wavelength.

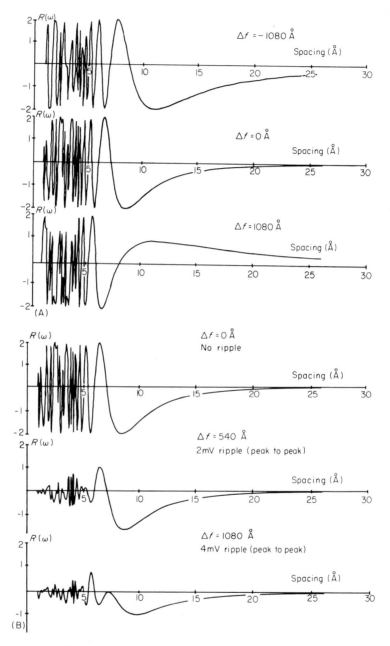

FIG. 2. (A) Phase contrast transfer function (79), $R(\omega) = -2\,\sin[2\pi W(\omega)/\lambda]$ calculated for a Siemens Ia objective lens of focal length 2.8 mm at 100 kV. The

For the electron microscope objective this is about 1.6 wavelengths at 100 kV for a reflection of 5 Å, and hundreds of wavelengths for a 2 Å reflection. It is clear that for reflections of 1–2 Å we have very large spherical aberration modulations on the transmitted wave. The amplitude or contrast of these frequency components is consequently modified in the image. [This can be expressed by a *contrast transfer function, T* (ω) *(56, 79)*.] The dynamic aberrations act to reduce the effectiveness of the higher spatial frequencies. The ultimate resolution of the objective in terms of diffracted waves is given by the limiting spatial frequencies, ω, for which $T(\omega)$ still has significant contrast in the image.

As already noted, in the electron microscope the resolution limitation is quite different from the light microscope due to the difference in wavelength. In the electron microscope, diffraction spectra are readily recorded to about 0.6 Å, i.e., all the way to the expected temperature factor cut-off. Hence, the aperture of the lens is not the problem. However, in the electron microscope lens without any objective aperture, there may be a significant reduction in the effective amplitude or contrast $T(\omega)$ in the final image of the higher spatial frequencies. There are several factors contributing to this. The first is the large number of wavelengths of spherical aberration introduced into the higher spatial frequencies. Assuming that the accelerating voltage and lens focal length are held sufficiently steady, there is now a problem due to lack of longitudinal coherence. If the coherence is not adequate, the phase of the diffracted waves will be randomized to some extent. The same effect occurs as a result of variations in the effective focal length due to fluctuations in the high voltage and objective lens current. The drastic effect of integrating only the objective supply fluctuations is shown in Fig. 2A and B. The phase contrast transfer function *(79)* is largely damped out, although asymmetry in this function leaves a small resultant effect.

The phase contrast transfer function $R(\omega)$ is a proportionality

---

function shows the phase distribution of spatial frequency components of the Fourier transform of the object, and indicates which spacings should be accentuated by a negative or positive phase contrast effect *(38a)*. (B) Same phase contrast transfer function with 2 mV of 60 cps ripple added. The phase oscillations are markedly damped out by this relatively small amount of AC ripple suggesting that the ripple diminishes contrast *(38a)*.

factor between phase distributions in the object plane and intensity distribution in the image plane. This is defined by:

$$R(\omega) = -2 \sin\left[2\pi W(\omega)/\lambda\right] \tag{10}$$

where ω is the spatial frequency, and $W(\omega)$ is derived from the wave aberration $W(\alpha)$, the phase shift of diffracted rays with respect to the nondiffracted ray. The latter is given by the Scherzer expression (105) which relates the phase shift due to spherical aberration to the phase shift due to defocus Δf:

$$\phi = -\frac{\pi}{2\lambda}\left(C_o\alpha^4 - 2\Delta f\alpha^2\right) \tag{11}$$

The transfer function $R(\omega)$ represents an operator unique to a particular lens which operates on the phase and amplitude of each diffracted ray passing through the lens. The defocus term Δf can be a combination of a static and dynamic focus changes due to supply instabilities.

Thus, an electron microscope without a back focal plane aperture behaves as if a semitransparent physical aperture exists in the back focal plane. Its opacity increases as a function of α^4 because the limited longtitudinal coherence of the beam causes the phase of wide angle scattered rays to assume random values with respect to the nondiffracted beam. In a sense, a physical objective aperture to cut off the 4–0.6 Å reflection appears unnecessary since these reflections are damped out by incoherence, AC ripple, and other supply fluctuations. However, inserting an objective aperture to cut out these spatial frequencies has a very obvious effect in increasing contrast, thus demonstrating that incoherent damping is by no means fully effective in suppressing them.

In the light of fluctuating phase shifts for wide angle spectra, it is worthwhile to reexamine the mechanism by which the objective aperture increases contrast. The Bragg diffraction pattern, in or near the back focal plane, looks sharp and clear out to the widest angle ($\sim$0.6 Å). Except in some microscopes where the intermediate lens aberrations interfere, the line width and intensities appear to be in agreement with those to be expected from the structure of the diffracting crystal. The back focal plane pattern appears identical to that obtained by general area diffraction without the use of a lens. However, the sharpness of the back focal plane intensity image is misleading with respect to phase shifts in the reflected cone. The phase shifts only became apparent on mixing the diffracted rays with the nondiffracted ray. The phase shifts in the 4.0–0.6 Å wide angle diffraction spectra cause modifications in the contrast of the image. The phase shifts represent the sums of the path differences due to spherical aberration (from a fraction of one wavelength up to hundreds of wavelengths) plus random and 60 cps ripple fluctuations. The lack of longtitudinal coherence for rays with large path differences also serves to dampen out the contrast effect of these rays. The net effect is to reduce the contrast of details of size 4 Å or less such that even in situations where the radius of the aberration disc of confusion is less than 4 Å, such detail is not resolved because of lack of contrast. This raises another discrepancy with the conventional Abbe approach to image formation. There is a lack of correspondence between the amplitude and spatial frequency of components in the back focal plane and the resolution of the image for an electron microscope. In light optics, aberration phase shifts are usually small and amplitudes of diffraction spectra affect the resolution. In electron optics only phase shifts are influential.

Even though the contrast effect of wide angle spectra is severely dampened there is still a large increase in image contrast on inserting a back focal plane aperture and removing them from the image. Hence, the effect of the aperture is simply to reduce the brightness of image structures; the result is similar to what would have been observed if the object had absorbed a significant fraction of the incident beam.

Without further detailed information about the frequency response, in terms of phase shifts of electron microscope lenses, it is impossible to say whether the static phase shift of spherical aberration, the dynamic phase shift of 60 cps ripple, and random supply instabilities, or the lack of longitudinal coherence are responsible for the present practial limitation of resolution for the electron microscopes to 3–4 Å. However, there is a possibility that estimates of the limiting resolution of present day pole pieces may be too pessimistic. To achieve 1 Å resolution it may not be necessary to shorten the focal length of microscope objective further to reduce aberrations, or to overcome spherical aberration by a correcting device (*107*). Instead it may require the further stabilization of high voltage and lens supplies and

the improvement of coherence of illumination by improved methods
of using thermal or field emission type of point cathodes. The success of the latter appears to require redesign of the microscope vacuum
system to allow efficient differential pumping of the filament area.

C. The Electron Microscope Image as a "Shadow Image"

When an electron microscope is operated with a back focal plane
aperture which cuts out reflection from 4 to 0.6 Å it is usually assumed,
because of the Abbe theory, that a resolution in the range of 4–0.6 Å
is unobtainable. It is often said that the only information about such
spacings is carried by the spectra that have been cut out by the
aperture. It is worthwhile reexamining this concept because of the
essential differences between electron scattering and light scattering,
and because thorough experimental tests of the resolution with and
without an aperture are not available. The effect of the aperture on
contrast shows that electron scattering has led to a significant amplitude modulation of the nondiffracted beam. The information carried by the diffracted rays is "mirror imaged" on the nondiffracted
beam. Here, it is carried on the lens axis as an amplitude modulation
with zero phase shift and no lens aberrations. To some degree, the
image may be a "shadow image." More exactly, it may be similar
to that of stained section in a light microscope which is only weakly
scattering, but strongly absorbing at certain object points. The
concept of electron imaging by amplitude modulation of the nondiffracted wave requires that the object detail is considerably greater
in size than the wavelength. This is, in fact, the case since at a resolution of 3 Å, spacings of 80 wavelengths are being resolved and even
at 1 Å, spacings of 27 wavelengths size are being resolved. The
practically attainable resolution is still many times the wavelength
for electrons, whereas for the light microscope it is of the order of the
wavelength. This represents another difference between the two techniques which needs to be taken into account on applying light optics
theory to the electron microscope.

The use of the objective aperture appears at first sight to be the
ideal imaging arrangement. The diffracted rays carrying the lens
aberrations are cut out. Their removal also serves to increase image

contrast; however, the approach is limited by Fraunhofer diffraction of the aperture and by other effects. Spectra having more than one wavelength of aberration (smaller than 4 Å spacing) are cut off by an aperture of about 50 μ diameter ($\alpha = 10^{-2}$) in most microscopes. The diffraction image from such an aperture has a size of $0.6\lambda/\alpha$ or $0.6 \times 0.037/10^{-2}$ Å or 2.2 Å. At the present time a number of claims are being made for a resolution of 2.0 Å or slightly less. In some cases an objective aperture is being used, in some cases not. One possible explanation for cases where unexpectedly good resolution is obtained with an aperture is that the Airy disc is not correctly calculated. The edges of the aperture are only illuminated by the weak diffracted rays, whereas the intense nondiffracted beam passes straight through. The advantage of tilted illumination (*27*) for improving lattice resolution is well known. If the principal reflection is placed on or near the axis of the lens, then spherical aberration will be minimized for diffracted rays associated with fine details. It will be imposed on small angle reflections which are now off axis, but these are associated with coarse specimen features. The fact that Dowell (*27*) found the technique to be limited in its effectiveness suggests that spherical aberration of the scattered wave is only partially limiting the resolution. The best lattice resolution yet reported is the resolution of the **220** planes of copper by Watanabe *et al.* (*117*) using a tilted beam (Table I). The **220** beam was admitted to the image, but it is not clear if a back focal plane aperture was used to admit the zero order and **220** reflection. However, Yada and Hibi (*120*) where able to come close to this resolution by resolving the 1.02 Å, **200** plane of gold using axial illumination. This was achieved by using a point cathode (*119*) in place of a normal hairpin filament. Thus, at this time there are conflicting viewpoints about the factors limiting resolution, and it appears likely that several factors are operative: supply instabilities, AC ripple, and spherical aberration and coherence. Clearly, critical experiments are required to assess these factors quantitatively. One question that needs to be definitely settled is whether a fine spacing (e.g., 2 Å) can still be resolved when the 2 Å reflection (and higher orders) are cut out by an objective aperture of suitable size. This would indicate the relative importance of the Abbe approach and the "shadow image" approach to electron imaging. This requires a lattice structure which is imaged

TABLE I

EXPERIMENTAL CONDITIONS LEADING TO HIGH RESOLUTION

Microscope	Voltage (kV)	Point filament	Tilted illumination	Objective focal length (mm)	Objective aperture	Observed lattice resolution (Å)	Observed point resolution (Å)	Test object	Modifications to the microscope	Reference
Hitachi HU-11B	100	+	−	2.1	200 μ	2.01	—	{200} in LiFl	—	121
Siemens, Ia	100	+	−	1.9	—	1.7	—	Graphite	Reduced AC ripple	62
Siemens, Ia	100	+	−	1.8	?	3.7	2.8–3.0	Biological	Reduced AC ripple	43
Hitachi HU-11B	100	+	−	2.1	200 μ	1.02	—	{200} Au	Also by strioscopy	120
Hitachi	80–100	+	−	2.1–2.5	?	3.81	—	MoO_3	?	66, 78
Siemens, Ia	100	+	+	2.8	Pass (III) (III)	3.2	—	{111} Ge	Extra good vacuum	97
Hitachi	80–100	−	+	2.1–2.5	?	1.27	—	{220} Ca	—	117

primarily as a result of Bragg diffraction and not due to phase contrast or Fresnel diffraction effects. In most cases, quantitative explanation of lattice images is incomplete. Organic crystals frequently show lattice images after the electron diffraction pattern has faded out. Presumably, the most satisfactory test specimen will be an inorganic crystal with a stable diffraction pattern and with the lattice image visible at focus.

III. Biological High Voltage (1 MeV) Electron Microscopy and Reduction of Specimen Damage

In discussing factors limiting the resolution we have assumed that the specimen will remain in a highly ordered state while being examined in the electron microscope. The delicate nature of hydrated biological objects makes this a most unlikely assumption. As indicated in the previous section a close examination of the factors limiting resolution may well suggest the way to reach 1 Å resolution in the next few years. However, biologists are no doubt wondering whether they can make use of an electron microscope resolving 1 Å when it comes available. There are indications that it will be impossible to maintain the specimen in a highly ordered state using microscopes operating at the conventional 40–100 kV accelerating voltage. This is indicated by the wide angle electron diffraction patterns of crystalline biological materials which indicate, with sensitivity, the degree of order of spacings in the range 4–1 Å. In most cases the diffraction pattern disappears in seconds and in many others no pattern is observed at all.

It is a striking fact that it has proved possible to produce only a few electron diffraction pictures of biological (nonmineral) objects. There is no essential difficulty in producing thin crystals and orientated layers in a suitable form for electron diffraction (less than 900 Å thickness at 100 kV). If the support film is made hydrophilic by glow discharge treatment in a vacuum evaporator, thin regular crystals of proteins, for example, can often be obtained by simply drying down a dilute solution onto the film. In all but a few cases where unusual bonding of the structure gives extra stabilization, diffraction patterns are not obtained even though morphologically the crystals look suitably ordered. These negative results are not improved by use of a tilting stage. The problem is not one of placing the crystals at the exact Bragg settings. The thin crystal layers lead to a relaxation in the Bragg condition such that each reciprocal lattice point is folded with the shape transform of a thin plate to give a reciprocal lattice spiked in a direction perpendicular to the thin layers. The nearly flat surface of the Ewald sphere of reflection can thus interact with the spiked reciprocal lattice net over a wide range of settings, often over more than 30° specimen tilt range.

Fiber orientated biological macromolecules have given useful indications of the disordering factors involved in electron microscopy. In some helical macromolecules the strength of the structure lies in internal hydrogen bonding. With others the structure is also dependent on hydrogen bonding with the external water structure. Examples of the first group are polybenzyl-L-glutamate (PBG) (*96*) and poly-riboadenylic acid (*94*) which give electron diffraction patterns in vacuum. Examples of the second group are DNA and RNA from which electron diffraction patterns cannot be obtained. The disordering in this case is partly due to dehydration and not beam radiation or thermal damage. This is readily proven by x-ray diffraction experiments on nucleic acids in various states of hydration. In the fully hydrated states, DNA fibers assume Type B form, with slight drying the still regular Type A form, with further drying the poorly ordered Type C form, and with further drying the diffraction pattern disappears. These effects of drying occur under exposure conditions such that the x-ray beam does not cause thermal or radiation damage.

However, for all organic crystalline materials (hydrated or not) the electron beam damage is very significant. In obtaining electron diffraction patterns from PBG and poly A, we found that cooling the specimen to about −10°C increased the lifetime of the diffraction pattern about two times. Kobayashi and Sakaoku (*77*) have noted a similar limited effect for the electron diffraction pattern of poly-ethylene crystals. Maintenance of the specimen temperature at 20°C increased the lifetime of the diffraction pattern, although lower temperatures gave no further improvement. It appears that about half (or less) of the beam damage is due to thermal damage, but the rest is due to radiation damage. Clearly, high resolution biological electron microscopy requires that the specimen receives the minimum electron beam bombardment in order to give the required electron diffraction pattern or image. Perhaps the least complicated measure to be taken is to reduce the size of the irradiated area to the bare minimum required. This raises the question of whether the double condenser system and its two condenser apertures are being used in optimal fashion in present day electron microscopy. To most electron microscopists the choice of apertures and setting of focal currents for the two condensers is largely guess work. It is surprising that the manufacturers have not carried out computer calculations to settle this point and to provide the microscopists with a table giving the best combination of apertures and lens current settings for every magnification. It appears that point cathodes can help to bring the irradiated area of the specimen down to a minimum; unfortunately, their effective operation requires an additional dynamic pumping system for the gun of the microscope that is not currently made available by the manufacturer.

Given that the diffraction pattern or image is being produced with the minimum irradiation of the specimen, it may be asked whether an image intensifier could reduce further the specimen damage occurring during recording of the diffraction pattern or image. Conventional screen illumination conditions for imaging [3×10^{-4} L or 3×10^{-11} A/cm² or 1.9 electrons/second/μ^2 at 100 kV according to Haine and Cosslett (*53*); or about 1 electron/second/μ^2 according to Valentine (*115*)] are already fairly close to the limitations of statistical noise producing "graininess" of the electron micrograph. This is obvious on exposing one plate to an electron beam (without specimen) to an optical density of 1.0 and another to daylight to the same density, and viewing the similarly developed plates with a hand lens. The electron exposed plate has a marked "graininess" due to statistical fluctuations or bunching of electrons. At first sight, reduction of the average image intensity by more than 10 times would seem to involve a significant loss of resolution due to increased electron noise. However, this is not quite the situation. First, the statistical noise depends on the total number of electrons required to obtain the image while the specimen damage depends on dose rate to some extent. An intense illumination for a short time may raise the specimen above a melting point or charring point while an equivalent prolonged exposure at low intensity may not. Second, the conditions required for satisfactory recording of electron diffraction patterns are entirely different from those required for final images. The resolution required for a diffraction pattern is much less; consequently more statistical electron noise can be tolerated. A study of the use of an RCA image intensifier for the recording of the electron diffraction patterns of crystalline polymers has been reported by Morrow and Horner (92). Using the image intensifier, a final screen intensity of only 4×10^{-5} L was used to give reasonably sharp electron diffraction patterns on the TV monitor of normal screen brightness. At 100 kV accelerating voltage, the lifetime of the diffraction pattern of crystals of polyethylene was now increased from 1 minute or less at normal screen intensities to more than 45 minutes. Hence, image intensification is an effective way of reducing the damage of organic materials in conventional microscopes when diffraction data is required, but it cannot be expected to be nearly as satisfactory if final image data for the intact specimen is also required.

A more direct approach to reducing specimen damage is to reduce the fraction of the incident radiation that deposits energy into the specimen by the process of inelastic scattering. This is predicted to occur at higher accelerating voltages (500–1000 kV) (1, 90). The general predictions have been confirmed by studying the lifetime of the diffraction patterns of polyethylene and other organic crystals at different accelerating voltages from 50 to 500 kV (76, 77). The cumulative dose (coulomb/cm²) to cause disorder and loss of the diffraction pattern was increased seven times in increasing the accelerating voltage from 50 kV to 500 kV. The experimentally observed increases in lifetime of diffraction patterns of organic crystals have not yet been reported, but Kobayashi's results suggest that an improvement of at least 10 times can be expected in going from 50 kV to 1000 kV. A moderate amount of specimen cooling might well increase this to about 20 times. However, it is likely that considerable variations will actually be observed with different types of organic crystals since the radiation damage is related, in part, to the ability of the molecules in the crystals to cross-link with one another.

From the very limited experimental data it might appear that the use of an image intensifier on a conventional microscope is a more effective (and much cheaper) way to reduce specimen damage than use of a high voltage microscope. However, the necessity that electron noise should not obscure fine detail severely limits the image intensifier for use with final images rather than diffraction patterns. In addition, it is much easier to construct specimen microchambers that keep the biological sample in a hydrated state at high voltages than at conventional voltages, because the device can have thickness dimensions increased about 10 times. A further advantage of the high voltage microscope is that the diffuse background of inelastic scatter is reduced with effectively an increase in signal/noise ratio. This improved ratio should not only allow the contrast stretching procedures required to bring the specimen contrast at 1 MeV up to the level at 100 kV, but also allow improved image processing at the higher voltages in order that fine detail can be accentuated in the processed picture. In spite of its great expense, the high voltage approach to specimen damage generally appears to provide a better solution to the problem, and should be used in combination with image intensifiers.

The choice of an optimum acceleration voltage of biological high voltage electron microscopy is a difficult matter. Heidenreich *(63, 64)* considers that for the out-of-focus phase contrast mode of imaging, the optimum voltage is only 200 kV. The decrease of phase contrast has been balanced against the decrease in specimen damage. However, this calculation does not take into account the improvement in signal-to-noise ratio in the image which allows contrast to be recorded. Assuming a decrease in inelastic scattering of about 10 times, it appears likely that S/N is increased about 10 times; it should be possible to restore the contrast by on-line computer processing of the image, electronic contrast stretching in an image intensifier TV system, or a combination of the two. In the final analysis, an experimental exploration of the possibilities will be required over a wide range of voltages. It appears resonable to limit the initial survey to 1 MeV because of the increased costs of high voltage generators and lenses above this value and the increased difficulties experienced with hard x-ray production.

Many technical developments are required to adapt commercially available 1 MeV electron microscopes to biological work. In making use of the objective aperture-type of amplitude contrast, several difficulties exist. First, the apertures are a source of hard x-rays and adequate shielding has to be provided. Second, the edges of current types of platinum apertures will be transparent to 1 MeV electrons. Third, the apertures must be decreased in size with each increase in voltage in proportion to the decrease in wavelength. Thus, straight bore thick apertures are required in the range of $5-10\,\mu$ at 1 MeV. These will require to be heated to eliminate contamination, although the contamination rate is much reduced at 1 MeV due to the decreased interaction with hydrocarbon vapor molecules.

The Toulouse group (31) constructed microchambers for examining hydrated specimens (living bacteria) at about 1 MeV. These were basically similar to those developed by Stoianova and Mikhailovskii (110) and Heide (60) for 100 kV accelerating voltage. The specimen (live bacterial cells) was trapped between two thin films. The sealed two-film chamber is much easier to operate at 1 MeV than at 100 kV because thicker sealing films can be used.

An alternative arrangement is the differentially pumped system developed in our own laboratory for 100 kV operation (Fig. 3). The wet specimen is supported on a grid and placed inside the small inner chamber. A flow of water vapor or helium saturated with water vapor flows over the specimen and escapes into the second chamber through $70\,\mu$ apertures. The second chamber is pumped by a high speed oil diffusion pump which has a liquid nitrogen cooled surface at its inlet to trap the water pumped off. The whole of the specimen chamber area is now pumped with a second larger oil diffusion pump so that a vacuum of 10^{-4} torr is maintained. The rest of the microscope is maintained at 10^{-4} to 10^{-5} torr.

As will be discussed in the section on contrast, it appears necessary to improve the high voltage and objective current stabilities. The most promising way of stabilizing the objective is probably to exchange the conventional objective for a superconducting lens which would have very much greater magnetic field stability. Recently, considerable progress has been made toward constructing practical superconducting lenses for use at 100 kV, but difficulties in finding a suitable high saturation permeability material for the container surrounding the niobum–tantulum ring or discs have limited progress in making superconducting lenses for use at 1 MeV. To obtain the desirable short focal length of 1–2 mm, it is necessary to produce

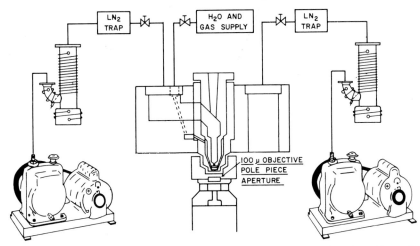

Fɪɢ. 3. Differentially pumped, wet specimen chamber developed at Roswell Park Memorial Institute. The specimen is placed on a grid between two platinum 70-μ apertures. Water vapor is admitted to the specimen space and flows out through the apertures. The water vapor is then pumped off by cryogenic and fast (900 L/second) oil diffusion pumps so that a normal vacuum is maintained in the microscope. Wet biological specimens can be maintained in various hydration states by this equipment and examined by electron diffraction and electron microscopy (*96a*).

an axial field distribution of sufficiently narrow half width. If the container material becomes saturated, the half width increases and the focal length increases. As a consequence, the spherical aberration also increases.

Special photographic plates have not yet been developed for use at 1 MeV. Thicker emulsions are required and perhaps the addition of colloidal metals to increase their stopping power. Most present day emulsions are completely penetrated by even 100 kV electrons. The photographic manufacturers have never attempted to match emulsion thickness to accelerating voltage.

The recent success of the RCA (United States), AEI (Great Britain), Hitachi (Japan), and JEOLCO (Japan) companies in making available 1 MeV electron microscopes for routine operation makes the question of their biological usefulness a timely one. These instruments are mainly developed on a commercial basis to meet the need of the metallurgist to examine thicker and more intact metal sections. As already discussed, the 1 MeV electron microscope is also strongly recommended for biological work mainly because of the need to reduce beam damage to delicate biological objects. It is natural for grant agencies, institutes, and universities to hesitate before spending a total of about 1 million dollars for 1 MeV electron microscopes and appropriate building; however, it should be emphasized that high voltage electron microscopy provides the most promising route towards 1 Å resolution of biological materials in their natural state. In spite of the high cost of the equipment and the necessity for a group of biologists to employ two engineers to assist in running the machine, the possible gains in terms of solving pressing medical and biological problems through the high resolution structure analysis approach appear to justify the cost and complexity. The technique of biological electron diffraction (95) is a very promising one if the specimens can be kept in the hydrated state. The major complication in applying the 1 MeV electron microscope to biological materials is the decrease in contrast at high voltage. To recover the contrast, it will be necessary to view a TV monitor (with or without computer processing). These are formidable complications for biologists attempting to advance electron microscopy to undertake, and it is regretable that a closer affiliation of biologists, physicists, and manufacturers has not been worked out. Electron microscopists have complained for more than 10 years about the backward design of electron microscopes, but with little visible effect on manufacturers. It seems incredible that one can still spend about $60,000 on an instrument which is packed full of solid state circuitry, and which still has the main inconveniences of instruments of 10 years ago. In most instruments no attention is paid to preventing the contamination of objective apertures even though this factor, second to specimen drift, is the most important one in obtaining consistent high resolution pictures. The design of the vacuum system of many current electron microscopes is difficult to understand. In some cases a very low speed mercury diffusion pump has been placed between a high speed oil diffusion pump and a mechanical pump, thus limiting the pump speed of the whole system and producing high contamination rates. In the light of present day vacuum technology (turbomolecular pumps, ion pumps, etc.), it would be an easy engineering task to provide microscopes with efficient pump speeds and low contamination. It is time for electron microscope manufacturers to consult with panels of experienced users and to provide instruments that are more up to date with respect to current engineering techniques.

Electron microscopy appears now to have reached a critical point in its development where considerable engineering and physical development is required in order to make further progress. It is unfortunate that institutes of electron optics were not set up to act as advisory centers, both for users and manufacturers in the design of advanced models of electron microscopes. The absence of a National Center for Electron Optics Research is particularly difficult to comprehend in view of the widespread benefits that should ensue from developing electron optics to many branches of pure and applied research and to industry. It is obvious that the problems of designing advanced forms of electron microscopes (both for physical and biological use) are related to those of design of electron beam welding and microetching machines, microprobe x-ray analysis equipment and also design of linear accelerators, plasma research equipment, radio tubes, klystrons, and magnetrons. A far sighted government should find it beneficial to invest in such an electron optics center in terms of the benefits to be expected for a wide range of electrical engineering industries.

Readers interested in discussion of the design of high voltage electron microscopes should refer to recent reports (*3, 17, 57, 103, 111*).

The "clarity" of high voltage diffraction patterns and images has often been noted. The chromatic aberration due to energy loss ΔE in the specimen is:

$$\Delta r = C_c \alpha \cdot \frac{\Delta E}{E} \qquad (12)$$

where Δr is the radius of the disc of confusion produced, α the scatter angle, E the energy of the incident beam, and C_c is the chromatic aberration coefficient. The specific inelastic energy losses, ΔE, for a given material are constant and independent of accelerating voltage; thus, the fraction $\Delta E/E$ is decreased by increased accelerating voltage. In addition, scatter angle α is also reduced; thus, chromatic aberration Δr is reduced. The resolution of dark field or strioscopic images is in part dependent on chromatic aberration, and in part on spherical aberration. It is found that such images are much sharper at 1 MeV compared to 100 kV (see Section IV). Hence, dark field and strioscopic methods provide a simple way to obtain high contrast images of weakly scattering biological objects in the 1 MeV electron microscope.

IV. New Methods of Obtaining Contrast

In the recent development of the electron microscope more attention has been paid to resolution than to contrast. However, for biological materials contrast is a key problem. The similar atomic constitution of different cell structures (membranes, ribosomes, chromosomes, etc.) results in very poor discrimination between them when viewed in the electron microscope as thin unstained specimens. This can be readily seen by surface spreading on distilled water an isolated membrane preparation and examining it on a grid without subsequent staining. Thin sections of glutaraldehyde-fixed cells (but without postfixation with osmium tetroxide) are almost featureless. Denser portions, such as nucleoli, can just be distinguished. The low contrast is especially a problem at true focus where the highest resolution can be obtained. In some cases underfocusing to produce out-of-focus phase contrast gives moderate contrast. Surprising contrast has been obtained in untreated bacteriophage particles by this method (14). In general, the back focal plane aperture method of imaging gives inadequate contrast on unstained specimens at focus. The contrast of positively and negatively stained or metal shadowed objects is entirely due to the metal. Since all techniques of metal contrasting appear damaging to the specimen and of inadequate resolution, it is appropriate to reexamine the imaging mechanism used in the electron microscope to see if more contrast can be obtained by a different imaging mode. The most satisfactory solution to this problem in the light microscope has turned out to be the in-focus phase contrast technique. The degree of phase contrast obtained is proportional to the fraction of the incident beam scattered by the specimen. This suggests that it is an appropriate approach in the electron microscope since, although organic materials are weak scatterers in relation to metals, the level is still significant. This is shown by the fact that electron diffraction patterns can be recorded from monolayers of hydrocarbon 20 Å thick at 40–100 kV acceleration voltages. We will discuss recent investigations of the electron phase contrast approach in this section. The in-focus phase contrast technique, where the scattered beam is retarded one quarter of a wavelength by a thin film phase plate or electrostatic field, will be compared with the out-of-focus phase contrast method where the scattered beam is retarded by underfocusing.

A second possible approach is the dark field and strioscopic techniques. A third approach is to stretch the contrast by image processing by computer.

Considerable use of low voltage electron microscopy has been made as a method to enhance contrast (*24, 118*). In the Philips EM-300, and earlier versions, the anode to cathode distance can be decreased so that illumination is not lost at low accelerating voltages and the microscope can be operated at 20 kV. Van Dorsten obtained a resolution of about 25 Å at 13 kV with considerably increased contrast. Fresnel fringe intensities were so weak that difficulty was experienced in compensating the astigmatism. Wilska (*118*), however, expects a resolution of less than 10 Å at 6 kV acceleration voltage and quotes the high contrast and small size of the instrument (power supplies, lens coils) as particular advantages. Electrostatic disturbances due to charging of the specimen and microscope parts close to the beam, the strong influence of stray magnetic fields and the necessity for very thin sections and objects proved difficulties at low voltage. However, in terms of the ultimate object of 1 Å microscopy of biological materials in their natural state, the low voltage approach seems contrary to the necessity to minimize the beam damage resulting from inelastic scattering since the ratio of inelastic to elastic cross sections is greater at low voltages than at conventional voltages.

The most elegant approach to low voltage electron microscopy has been recently provided by Heinemann and Mollenstedt (*65*). An electrostatic lens was used to decelerate 40 kV electrons to a few kV at the level of the specimen placed at the center of the lens. The bottom part of the lens served to reaccelerate the electrons to 40 kV. The top part of the lens acted as a condenser and the bottom as an objective. A theoretical resolution of 10 Å is expected. High contrast photographs at 5 kV were obtained.

The definition of contrast is a matter of some difficulty. Ultimately, the definition has to relate the physiological response of the eye in viewing a photographic plate, print, or TV monitor. In most quantitative discussions it is defined as the difference in intensity between the structure and background divided by the background intensity:

$$C = \frac{|\Delta I|}{I} \tag{13}$$

This definition takes into account the fact that for the ability of the eye to discriminate a *minimal* intensity difference is related to the background intensity. The eye appears to "saturate" when a bright field is presented to it and small intensity differences cannot be distinguished. Since the end result of all electron microscope work is a direct visualization of the material under study, it is unfortunate that a better physiological understanding of contrast discrimination by the eye is not available. The definition of Eq. (13) has other advantages if contrast is being studied by microdensitometer scanning of photographic plates. Since photographic emulsions are almost perfect recorders for electrons (*115*), approximately one electron activates one silver halide crystal, whereas about 100 light photons are required per crystal. The emulsion density is proportional to electron intensity. Based on this fact it can readily be shown that contrast, defined by Eq. (13), is independent of emulsion sensitivity or rate of development of the plates. However, it should be pointed out that the photographic densitometric method of studying contrast is limited in accuracy and range of measurable contrast. In practice, it is more accurate and convenient to use a small Faraday cage to record structure and background intensities in the final image plane.

The recent introduction of densely packed arrays of several thousand light detectors (*36*) raises the question of whether similar arrays of solid state electron detectors could be used in the electron microscope in place of the fluorescent screen to give an image on a TV monitor. Such a detector array would combine the total picture integration ability of photographic plates with the wider range of intensity response of solid state detectors over that of the photographic plates.

Visibility of fine detail depends on size, contrast, brightness, and the length of time the retina is exposed to the image (*87*). In scanning an electron microscope image, either on a fluorescent screen or a TV monitor screen, the detection of structures of minimal contrast depends on having sufficiently high magnification and the optimum screen brightness. The size of the image and brightness of the field will also be dependent on the aperture and magnification of any binocular viewer used. In order to exceed the fluorescent screen granularity, the image detail should have a size of about 0.4 mm. According to Agar's results, as discussed by Haine and Cosslett (*53*), a screen brightness of 3 mL allows 0.4 mm detail with 20% contrast to be seen, whereas 0.4 mm, 40% contrast detail can be seen at a screen brightness of about 0.08 mL i.e., a brightness reduced by about 38 times. These results suggest that the contrast–brightness relationship is more complex than given by Eq. (13) and requires further investigation.

It should be noted that "instrumental contrast" is different from "eye contrast." A Faraday cage detector and also the photographic plate within a limited range can detect the same minimal contrast over a range of background brightness. The contrast given by densitometer tracing across a photographic plate may differ from that experienced by the eye on direct viewing of the plate.

A likely physiological contrast–brightness relationship would be one based on a logarithmic subjective response to stimuli (Weber-Fechner law). The response R is related to the intensity of the stimulus (I) by:

$$R = K \log_{10} I \tag{14}$$

so the intensity and response for a particle are I_P and R_P, and for the background R_B and I_B, then the contrast should be related to the difference in the two responses:

$$C = R_P - R_B = K \log_{10} I_P/I_B \tag{15}$$

in which the visual effectiveness of an instrumental contrast I_P/I_B is proportional to $\log_{10} I_P/I_B$.

The contrast definition of Eq. (13) has its limitations for dark field and strioscopic studies. In a perfect dark field situation a minimal detectable brightness of a dark field image would have infinite contrast since the background would have zero intensity. Some authors (*33*) in discussing dark field situations have used the ratio of the difference between the intensity of the structure I_S, and the background I_B divided by the average:

$$C' = \frac{I_\mathrm{S} - I_\mathrm{B}}{(I_\mathrm{S} + I_\mathrm{B})/2} \tag{16}$$

or

$$C' = 2\Delta I/(I_\mathrm{S} + I_\mathrm{B}) \tag{17}$$

Whether the contrast is accentuated by converting phase shifts of scattering to amplitude differences (phase contrast), reducing the background wave intensity (dark field and strioscopic methods), or electronic or computer contrast stretching procedures, the degree in which this can be successful depends on the signal to noise ratio (S/N). In the final image the intensity difference between the structure and the background must be compared to the random fluctuations in the background level of intensity. The background intensity fluctuations are due to the relatively small number of electrons arriving at each element of the picture. From Poisson's relationship the standard deviation or noise of the background is $N^{1/2}$, where N is the average number of electrons arriving per second at a single picture element. To this must be added the static, but random variations in intensity across the image due to structure in the specimen support film. In addition, the intensity over the structure is made up of the result of interference of the background wave with coherently or elastically scattered waves and incoherent or inelastic scattered component. The incoherent scatter has the effect of producing a background fog which obscures small contrast levels. The reduced inelastic scatter component obtained in the high voltage electron microscope gives an improvement in the S/N ratio for low contrast levels.

In the following some of the possibilities for amplitude or phase shifting, the scattered and nonscattered beams in the back focal plane of the objective will be discussed.

A. Strioscopic Dark Field Technique

In the dark field technique an aperture is placed in the back focal plane of the objective and adjusted so that the nondiffracted rays are stopped, but particular diffracted rays are passed. When the nondiffracted rays are stopped, but *all* diffracted rays passed, the technique is termed "strioscopy." In practice, a fine probe or beam stop is used just larger than the width of the nondiffracted beam. The strioscopic method attempts to make use of the maximum intensity of scattered radiation from the structure of interest and then to enhance its contrast by reducing the background radiation to as close to zero as possible. The contrast is reversed, i.e., highly scattering portions of the specimen appear bright against a dark background. As will be discussed later, not all the diffracted radiation can be used, in practice, if good resolution is required in the image. In addition, there are possibilities for strioscopy with tilted and nontilted beams.

It has been recognized for a long time that dark field techniques in general, and the strioscopic technique in particular, are capable of showing up with high contrast structural details which have low contrast in a bright field image. In the light microscope (*15*) the large scatter angles make it possible to produce dark field without using back focal plane apertures or stops. The special condenser used causes the light to strike the object tangentually so that only scattered light enters the objective.

The terminology describing different modes of dark field and bright field imaging is somewhat confusing [for reviews of the different techniques see Zernicke (*123, 124*), Françon (*47*), and Martin (*89*)]. The main bright field methods follow.

(1) Schlieren or Foucault method (*113, 114*): A straight edge beam stop is used to cut off half the diffraction field, but allows the central beam to pass. The technique is related to the Foucault knife-edge test method of measuring aberrations. The method has been used, in particular, to demonstrate separate layers of materials of different refractive index in the sample tube during ultracentrifugation. A straight edge cuts off the rays deviating from the main beam because of refractive index differences, so that the image of the column of liquid in the cell shows dark bands representing interfaces. The relief effect observed by schlieren light microscopy is discussed by Zernicke (*123, 124*). Also, a technique where the schlieren effect is obtained by reversing the phase of one half of the diffracted rays is described. The method appears not to have been explored in the electron microscope, although the use of an off-center aperture that just passes the central beam and most of one half of the diffraction pattern should give a similar effect.

(2) Central back focal plane aperture method: This is the common method of obtaining amplitude contrast in the electron microscope. The method has been used with the earliest models of electron microscopes. Von Boersch (8, 9) first used a back focal plane aperture and observed enhanced contrast and dark field effects.

The main dark field methods are:

(1) Off-center aperture ("dark field"): A back focal plane aperture of appropriate size is moved off center so as to stop the central beam and to pass selected diffracted beams. The image is reversed in contrast and emphasizes periodicities corresponding to the selected spectra. The method has been widely used in electron microscopy for viewing crystals and crystal defects (54, 55, 68, 116). The aperture is either moved off the axis of the objective lens or else the aperture is kept centered and the beam entering the objective lens is tilted. High resolution (12 Å) dark field microscopy can be achieved by tilting the beam to reduce the effects of spherical aberration and by reducing contamination of the crystals (51, 98). Hirsch et al. (68) reported a resolution of 12 Å on a Moiré pattern from overlapping gold films using tilted illumination.

Dark field methods provide one of the easiest ways to obtain a high degree of contrast. They have not been sufficiently well studied for enhancing the contrast of biological materials. Dark field methods seem particularly indicated for high voltage (1 MeV) electron microscopy of biological materials. However, their usefulness depends on the available resolution. A reexamination of all the factors leading to low resolution in dark field images would seem desirable.

(2) Central beam stop (strioscopic) method (28): A small circular beam stop (or a fine wire) is placed to obstruct the central beam and to pass most or all of the diffracted rays. The term strioscopy [L striae (filament)] relates to the use of the method to see small transparent filaments (89). However, the term has also been applied to the schlieren method where liquids show fine lines indicating refractive index changes (15). Essentially the use of a dark field condenser in the light microscope gives the strioscopic condition, since the cone of the nondiffracted beams is suppressed, and the cone of the diffracted rays passed. In the Spierer lens (106), a small mirror is placed close to the back focal plane of the objective to reflect back the non-

diffracted beam out of the image. The first investigators to use the method in the electron microscope were Fert and Faget (*45*). They used a slit type condenser aperture and a metallized fiber in the back focal plane of the objective and reported high contrast, dark field images of zinc oxide crystals and diatoms. The problem of low contrast of organic and biological objects in the high voltage (around 1 MeV) electron microscope stimulated further interest in the strioscopic method. Using the CNRS high voltage microscope at Toulouse, the French workers Dupouy *et al.* (*33*) and Dupouy and Perrier (*30*) used a 6 μ wire central beam stop to enhance the contrast of biological objects. A comparison was made between the strioscopic images obtained at 75 kV and 100 kV accelerating voltage and 1 MeV. Dupouy *et al.* (*34*) calculated that the contrast of the strioscopic method is six times greater for graphite than the conventional back focal plane aperture method at 75 kV accelerating voltage, and 7–20 times greater at an accelerating voltage of 1 MeV. At 1 MeV the thinnest film having minimal detectable contrast (5%) was calculated to be 130 Å by using a back focal plane aperture and 13 Å by the strioscopic method. Yada and Hibi (*120*) were able to resolve the 1.02 Å spacing of gold using a strioscopic technique in conjunction with a point filament.

In recent work (*70a*), Johnson and Parsons examined the potential of the strioscopic technique for electron microscopy of unstained biological material. Strioscopic beam stops were constructed by microwelding fine platinum wires (10–20 μ) onto platinum apertures of hole size from 30 to 750 μ (Fig. 4). The contrast effect of these beam stops were examined both in a Siemens Elmiskop Ia electron microscope and a Philips EM-300 microscope. The effect of placing the beam stop exactly in the back focal plane and in position above or below it was investigated. In a modified Elmiskop Ia (*70*) an additional intermediate lens allowed the position of the back focal plane with respect to the beam stop to be varied. In Fig. 4, the form of the strioscopic beam stop placed in the back focal plane of the Philips EM-300 objective lens is shown together with the diffraction pattern of molybdenum trioxide in the same back focal plane. The contrast reversing and enhancing ability of such a beam stop on several images is shown in Figs. 5–7. In Fig. 5 it should be noted that the thin carbon

FIG. 4. Strioscopic beam stop, consisting of a 30-μ heated platinum wire, is shown in the back focal plane of a Siemens Elmiskop Ia electron microscope. The diffraction pattern, focused in the same plane as the beam stop, is that of MoO_3 (No. 2998).

film shows much greater contrast in the strioscopic image suggesting that contrast is dependent on chromatic aberration.

In these studies an attempt has been made to better define the resolution limitations of the strioscopic technique. Crystal lattices and Moiré patterns were used as test objects, and the resolution determined with different sizes of aperture supporting the beam stop. Thus, the intensity distribution of spatial frequencies having particular static and dynamic aberrations were added to the image by varying both the type of crystal and the size of the support aperture. The extent to which spherical aberration contributes to the loss of resolution was also tested by tilting the beam so that the most intense diffracted beams lay along the lens axis. In addition, the effect of chromatic aberration was studied by correlating thickness of the crystal with observed strioscopic resolution.

It was concluded that strioscopy, using a particular size of support aperture, has advantage over, the off-center aperture dark field technique. The aberrations introduced into the image could be kept at a minimum because of their peripheral cut off by the aperture. In addition, the marked stigmatic effect of dirt on off-axis apertures is avoided with the strioscopic method since the central beam strip is fine enough to be kept heated by the electron beam. The intensity of the dark field image could be made brighter by the strioscopic technique. It was found that peripheral cut-off of the highly aberrant diffracted rays made tilting of the beam unnecessary.

B. In-Focus Phase Contrast Electron Microscopy

The phase contrast light microscope (*123, 124*) has proved to be an invaluable technique for examining unstained biological materials (*4, 6, 10, 47*). In transparent, only slightly absorbing objects that give a significant distribution of scattered or refracted light, the method gives much greater contrast than the bright field method. However, the theory and explanation of the phase contrast effect is often inadequately treated in the literature. Frequently, it is emphasized that phase contrast microscopy is similar to interference microscopy. The phase contrast microscope has often been regarded as an imperfect interference microscope where the beam transmitted

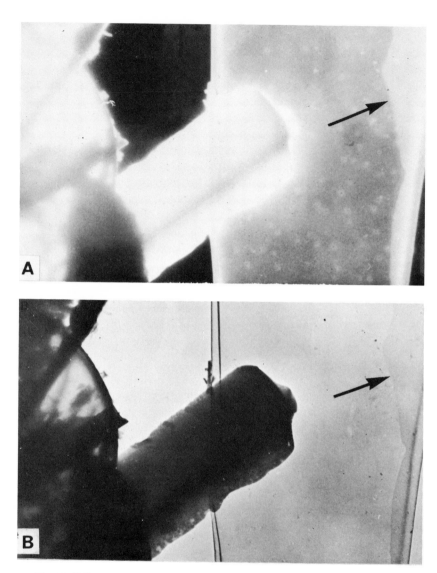

Fig. 5. Celite (diatomaceous earth) shown in dark field and bright field. (A) Strioscopic dark field, 10-μ platinum wire or a 750-μ platinum aperture (magnification 6700$\times$, No. 1782). (B) Bright field (no aperture). Note that the contrast of the fold in the 50–100 Å carbon field on the right is much greater for the strioscopic image (magnification 6700$\times$, No. 1783).

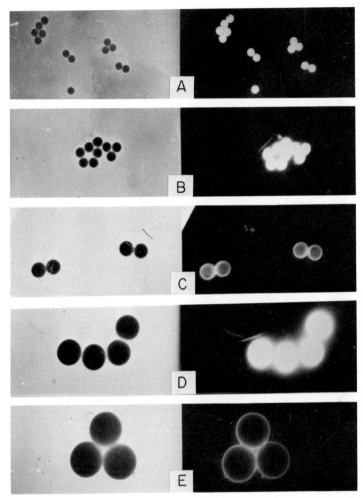

Fig. 6. Polystyrene latex spheres of diameters (A) 2540 Å, (B) 3570 Å, (C) 5570 Å, (D) 7960 Å, and (E) 11,000 Å shown by bright field (100 kV) without a back focal plane aperture (left and by strioscopy, with a 10-μ wire on a 750-μ aperture (right).

by the specimen is imperfectly separated from the reference beam. This approach obscures the fact that the phase contrast microscope depends entirely on the specimen scattering or refracting light out of the direct beam while the interference microscope does not (see

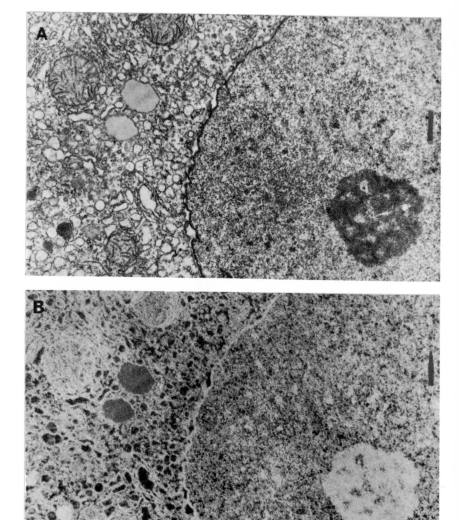

Fɪɢ. 7. Uranyl and lead stained, osmium tetroxide fixed, Epon embedded rat liver. Thin section on a bare 400 mesh gird. (A) Bright field with central 10-μ back focal plane objective aperture (No. 3346). (B) Dark field with the 10-μ aperture centered on the intense reflection due to the section stain (No. 3344). Magnification is 21,000$\times$.

Fig. 11). The essential operation performed by the phase microscope is to change the phase difference between the deviated and nondeviated beams by insertion of a thin film phase plate so that the maximum intensity difference or contrast is produced in the image.

The literature contains theoretical treatments of phase contrast microscopy which do not consider the angular distribution of deviated light. Such treatments cannot be quantitatively correct for a phase contrast microscope, but may be correct for an interference microscope. A complex transmission function is often used:

$$F(x,y) = A(x,y)e^{-\phi(x,y)} \qquad (18)$$

as the starting point for phase contrast theory and the Fourier transform of this used as equivalent to the diffraction pattern of the object existing in the back focal plane of the lens. However, the Fourier transform is actually related to the distribution of scattering points in the object and not as directly to the amplitude and phase changes in a nondiffracted beam traversing the object. The back focal plane diffraction pattern can be calculated by the normal x-ray crystallographic procedure, but needs to be modified by the lens optical transfer function. The final image is the transform of the modified diffraction pattern. The phase of each diffracted beam will be the transverse crystallographic or periodicity controlled phase, plus the appropriate longitudinal phase aberration produced by the lens. The contrast of the final image can now be modified by retarding either the diffracted or nondiffracted rays by introducing a thin film phase plate.

The maximum contrast is obtained when the scattered light is either in antiphase (diffracted beam retarded by π or $\lambda/2$) to give positive phase contrast where scattering objects appear dark against a bright background, or in phase (nondiffracted beam retarded by π or $\lambda/2$) to give negative phase contrast where scattering objects appear bright against a dark background. Experimentally, it is found that maximum contrast is achieved by a phase plate that produces a retardation of $\pi/2$ or $\lambda/4$ so the scattered wave must already be retarded by $\pi/2$ or $\lambda/4$. The explanation of this phase shift or scattering has been unsatisfactorily explained in the past using diagrams showing the background wave and the particle wave as having the same amplitude, but with a small phase difference. The phase difference is taken to be the optical path difference through the particle (P-wave), due to its thickness and refractive index, compared with the background wave (B) (Fig. 8a). As already pointed out, this concept is not valid since the phase shift in a phase contrast microscope is obtained by producing a relative shift of the deviated and nondeviated beams. Optical path differences in the forward direction produce no effect since the diffracted rays experience the same difference. The vector diagram is closed by a vector which is usually identified with the deviated wave (Fig. 8a, b). This is illogical since the amplitude and phase shift of the scattered radiation is unrelated to the amplitude and phase of the forward directed particle wave. In the case of light scattering it must be concluded that a phase shift of $\pi/2$ occurs. As pointed out by Fresnel, the radiation scattered from the edge of an aperture must be assumed to be phase shifted by $\pi/2$ in order to account for the observed diffraction pattern intensity distribution. It should be emphasized that phase shift of scattering is likely to hold generally for scattering of electromagnetic radiation, but not necessarily for scattering of particle waves.

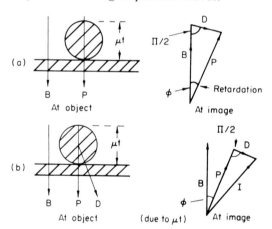

FIG. 8. (a) Refractive index or optical path difference concept of phase contrast. The vector sum of the particle wave (P) (which has been phase shifted because of an optical path difference) and a diffracted wave (D) is considered to equal the background wave (B). (b) Scattering concept of phase contrast. The diffracted wave (D) undergoes nearly identical optical path differences through the particle to the nondiffracted wave (P). The amplitude of P may be significantly less than B. The P wave and the D wave combine to give a resultant wave I for a point in the image of the particle. The scattered D wave underwent a phase shift of $\pi/2$.

In considering an in-focus electron phase contrast technique the above account shows that the critical quantity required to be known are the phase shifts of electron scattering. There is reason to suspect a difference from the electromagnetic radiation case. Secondary radiation is not involved in elastic electron scattering. The wave associated with each particle is modified by the potential field of the positive nuclear charge (*13*). Although some authors (*53*) have assumed that the deflection of an electron wave packet by nuclear charge involves a $\pi/2$ phase change, there is little theoretical justification for this. According to Zeitler (*122*), the first Born approximation gives a constant phase shift of $\pi/2$ when applied to single atoms, but a more exact theory shows an additional phase shift to occur. This suggests that even an aberration-free microscope behaves to some degree as a phase contrast microscope.

Recent calculations of Cox and Bonham (*22*) indicate that the phase shift on elastic scattering is variable with accelerating voltage and atomic number. According to Cox and Bonham (*22*), for the light elements of biological materials and accelerating voltages of 50–100 kV, the phase shift is small and possibly negligible. If this is the case, then a phase plate will be required to shift the diffracted electron wave by π, or some value just different from π, in order to achieve electron phase contrast. Depending on the value of the periodicity required to be enhanced in contrast, there may be a significant retardation due to lens aberrations (particularly spherical aberration and astigmatism). This is likely to amount to π radians for a spacing of about 3 Å and to a larger value for smaller spacings. If the spacing of interest is larger than 4 Å, the aberration phase shift (at focus) can be ignored.

The phase shifting device is a potential field of a few volts. This could be produced by a microelectrode arrangement in which the central beam or diffracted beam is retarded. A simpler method is to make use of the positive inner potential of thin films. The inner potential is the average residual charge existing in all solids as a result of incomplete screening of the positive nuclear charge by negative orbital electrons. For carbon at 100 kV accelerating voltage (V), the inner potential (E) is about 10 volts. The refractive index is given by:

$$\mu \simeq (1 + E/V)^{\frac{1}{2}} \simeq 1 + \tfrac{1}{2}(E/V) \tag{19}$$

and for the optical path difference (μx) to equal $\lambda/4$ or $\lambda/2$ the thickness, x is 185 Å or 370 Å. Beryllium films have the advantage of higher inner potential and smaller density than carbon. Even at 100 kV accelerating voltage the required thicknesses of phase plates are practical ones below the critical "Bremsdicke" value (11) of 11 $\mu g/cm^2$ for carbon or a thickness of about 550 Å. The Bremsdicke value gives the thickness below which only single inelastic scattering events occur.

Following suggestions by Boersch and Gabor, the first experimental attempt to obtain in-focus electron phase contrast was carried out by Agar *et al.* (1). They assumed that a phase delay of $\pi/2$ was required and used a Formvar film estimated to give this retardation. Holes of 1 μ diameter were obtained by pushing a glass fiber through the film. The film was placed exactly in the back focal plane and centered so that the nondiffracted beam passed through the 1 μ hole. It was found necessary to reduce the intensity of the focused beam spot in the back focal plane in order to minimize damage to the Formvar film. This lead to difficulties in focusing and photographing the image. No conclusion was reached about the efficiency of the system in enhancing contrast. The objects used were Formvar replicas of lightly etched metals.

Locquin (84–86) attempted to use a charged diaphragm to produce a relative retardation of scattered and nonscattered beams.

More detailed theoretical and experimental work on electron phase contrast was carried out by Kanaya *et al.* (72, 73) and Kanaya and Kawakatsu (71). These workers used collodion films containing holes of 5–50 μ diameter. A hole was centered on the nonscattered beam in the back focal plane so that the scattered beam only was retarded by passage through the film. Tissue sections and MgO crystals were examined using first a 50 μ objective aperture and then the collodion phase plate. The published photographs show contrast to be enhanced by the phase plate by several times. Although the authors emphasize that they expect "microcontrast" to be enhanced rather than "macrocontrast," it is impressive that both a tissue section and MgO crystals showed *macro*contrast enhancement of objects several microns in diameter. Edge fringes, characteristic of light phase contrast, were not reported. Experiments were also carried out with a charged conducting film to modify the phase of the diffracted beam.

Faget *et al.* (*39, 40*), Fert (*44*), and Fert *et al.* (*46*) made considerable progress in overcoming some of the technical difficulties in obtaining electron phase contrast. They increased the coherency of the beam by using a slit type anode. The back focal plane of the objective was made more accessible by running the intermediate at a low current and projecting the image of the back focal plane into a phase plate chamber. A second intermediate, together with the projector, then gave subsequent stages of magnification. The thin film phase plate was supported in a heated holder (250°C) to minimize contamination of the phase plate. The phase plate was of slit or trough form to suit the slitlike beam used. Carbon was shadowed over a fine wire to produce a trough of the required width and depth. Rotation coils were fitted to the condenser so that the beam could be aligned parallel to the phase plate. These workers used a carbon phase plate of step thickness 180 Å (it was assumed that a retardation of $\lambda/4$ was required). The width of the phase plate required to phase change the reflection associated with a particular spacing was calculated. The scatter angle for a spacing of d Å was λ/d (from Bragg's law applied to small angles). For parallel illumination the diffraction pattern occurs in the back focal plane (focal length $= f$). Hence, the reflection of interest occurs at a distance of $f\lambda/d$ from the lens axis. This is very approximate because of the strong curvature of the electron paths in this region. For a wavelength of 0.04 Å, Fert (*44*) calculated the following widths of the phase plate trough: 10 Å to 12 μ, 100 Å to 1.2 μ, and 1000 Å to 0.12 μ. Clearly the actual beam diameter, the focal length and accelerating voltage are parameters that require to be optimized and these will be discussed later. The French group have published pictures showing enhanced contrast (both macrocontrast and microcontrast) of carbon films with the phase plate in position. The phase pictures showed a fringe around the film which was interpretated as similar to that observed in light phase microscopy. The question of the scale of the contrast will be discussed later. Fernández-Morán (*43*) published high resolution micrographs where a phase plate has been used, but the details of the technique have not been published yet.

The current work in our own laboratory on developing an in-focus phase contrast system has developed further the technique explored by the French group. An additional intermediate lens was fitted to

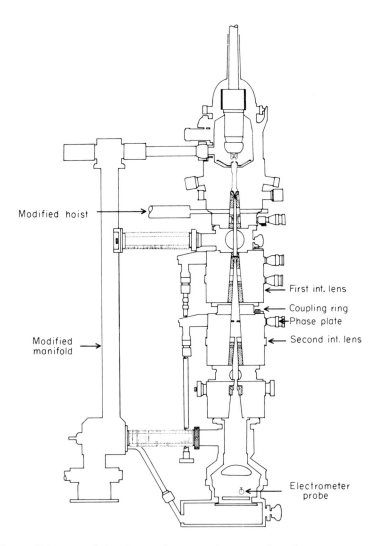

Fɪɢ. 9. Diagram of four-stage electron microscope for phase contrast and strioscopic experiments. The first intermediate lens is used to project the back focal plane of the objective into a more accessible space. This space accommodates the phase plate holder and an anticontamination device. The second intermediate lens, together with the projector lens, supplies a normal range of microscope magnification. Quantitative estimates of contrast are made by an electrometer probe.

a Siemens Elmiskop Ia microscope (Fig. 9) (*70*). The first lens was used to project the back focal plane of the objective into a phase plate chamber and the second used, together with the projector, to give a normal range of magnification. However, considerable simplification of the arrangement has been achieved by avoiding the use of slit-shaped beams and phase plates. A point cathode or hairpin cathode is used in conjunction with phase plates with circular holes. The phase plate is protected from contamination by a liquid nitrogen cooled metal enclosure (Fig. 10). The intensity distribution

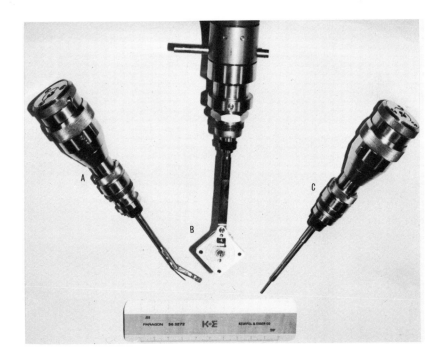

Fig. 10. Phase plate holders and anticontamination device. The phase plate consists of one square of a 100 mesh screen covered with a carbon film. A 5–15 μ hole has been drilled in the film by use of a microscope laser. This hole is aligned to pass the central beam of the object diffraction pattern. The phase plate is supported by the holder (A) on the left and positioned with the aid of the probe (C) on the right. The anticontamination device (B), in the center, prevents a change in path difference between the scattered and nonscattered beam due to deposition of contamination.

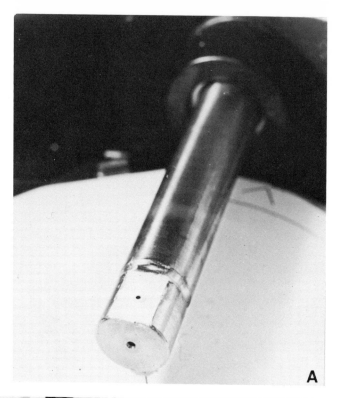

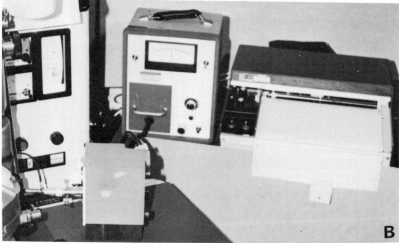

FIG. 11. (A) Micro-Faraday cage arrangement for accurate measurement of image intensity and contrast. The object is moved so that the intensity over a

or contrast was measured by moving the specimen so that different portions of the image fell on the opening of a small Faraday cage (Fig. 11). This proved a more accurate technique than microdensitometry of plates. The phase plates were made from carbon films of calibrated thickness. A technique of measuring thickness by electron microscopy of transverse sections of carbon films, embedded in Epon, cut with a diamond knife, has been developed by Moretz *et al.* (*91*). The direct measurements of thickness are used to calibrate a microdensitometer estimate of optical density of the carbon film. With this technique, thickness in the range 50–500 Å could be measured to within 10 Å.

It was found that 100 mesh specimen grids did not introduce significant aperture contrast into the image and these were used to support the carbon film. The phase plate consisted of one square of the mesh. Clean round holes of 1–50 μ could be formed in the center of one square by the use of a microscope laser (kindly made available by the American Optical Company, Buffalo). In positive phase contrast experiments the contrast of 880 Å polystyrene latex spheres and carbon films of calibrated thickness was measured with the phase plate in position, out of position, and also with a standard 50 μ objective aperture in position. The phase plate contrast relative to 50 μ aperture contrast and "no-aperture" contrast was measured for different thicknesses of phase plate and different hole sizes in the phase plate. Insertion of the phase caused a significant increase in macrocontrast (contrast of 880 Å latex spheres) as well as an increase in microcontrast as discussed below.

Negative phase contrast could be produced by using a thin (50 Å) carbon film in the actual or projected back focal plane. The back focal plane beam spot was used to build up a contamination spot of the same shape and size as the beam. Contrast measurements, while this was in progress, showed oscillating, maximum and minimum contrast levels. Presumably, the central beam was retarded successively by the critical values to produce maximum and minimum contrast as the phase plate increased in thickness (Fig. 12).

---

particle is first measured and then the intensity over the background or supporting film. (B) The intensity recording arrangement. The voltage developed on the probe is first preamplified and then fed to a Victoreen vibrating reed electrometer. From there it is passed to a strip chart recorder.

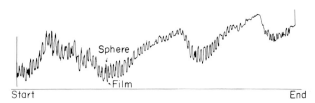

FIG. 12. A record of in-focus negative phase contrast of a single 880 Å poly-styrene latex sphere supported on a 50–100 Å carbon film. The contamination rate in the region of the phase plate was increased by the introduction of oil vapor. The central beam of the diffraction pattern then built up an increasing thickness of carbon over its area with only slight contamination in the region of the diffracted beams. The central beam was thus phase retarded by increasing amounts with respect to the diffracted beams. The Faraday cage intensity record, taken over 20 minutes, shows the intensity over the sphere and the intensity over the film. (The specimen was moved to and fro so that first the sphere image and then the film image were centered over the aperture on the Faraday cage.) The intensity difference divided by the film intensity gives the contrast. The record shows a marked in-focus phase contrast effect with maxima and minima. These, presumably, corresponded to thicknesses of the contamina-tion spot which give the critical retardations to produce phase contrast (70).

This work, in its current stage of development, has led to two main conclusions: first, that phase shifting the scattered beam with respect to the nonscattered beam produces macrocontrast as well as microcontrast. The explanation of this will require further study of the effect of phase shifting particular spatial frequencies in the back focal plane image. Second, the effect of placing a phase shifting device in the wide angle spectra region can only be explained if a better knowledge of the static and dynamic phase shifts associated with the wide angle region are determined.

Macrocontrast changes in contrast are frequently observed by the electron microscopist. On underfocusing he observes that large objects, microns in diameter, become more contrasty. In order to study the macrocontrast of 880 Å polystyrene latex spheres further, we meas-ured the contrast changes occurring on defocusing (70). A distinct minimum of contrast was observed at an underfocus of 6 μ. This ob-servation does not fit the Scherzer equation already discussed [Eq. (11)] which would predict many oscillations in contrast before a defocus of 6 μ is reached. This will be discussed further in relation to

out-of-focus phase contrast electron microscopy. However, it should
be pointed out that aside from the question of oscillations in contrast
due to interaction between a phase retardation due to spherical aberration and a phase advance due to underfocus, the general form of
the overall curve for change in contrast with defocus had not been
adequately explained. This will be discussed further in Section IV,C.

An electron phase contrast microscope would be expected to show
similar contrast characteristics to a light phase microscope. The
incomplete separation of the phase shifted and nonphase shifted
beams leads to a limitation in the area of the image over which
structures show enhanced contrast and also to a fringe at the edge of
objects [defective Zernicke condition (4)]. A similar effect can be

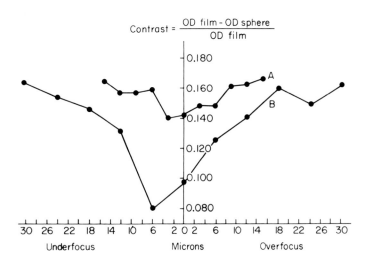

Fig. 13. Contrast changes for a single 880 Å polystyrene latex sphere as a
result of defocus of a Siemens Elmiskop Ia electron microscope (80 kV) using
a 50-μ objective aperture (A) and no aperture (B). Without the aperture, the
wide angle diffracted beams contributed to the image and a surprising contrast effect is shown for this *large* (880 Å) object. However, the well defined
minimum at an underfocus of 6 μ cannot yet be explained quantitatively as
due to a combined phase shift as a result of spherical aberration and defocus.
When the wide angle pattern is cut out by a 50-μ aperture, most of the contrast
effect is removed. The effect may be due to separation of dark and bright field
images as a result of spherical aberration.

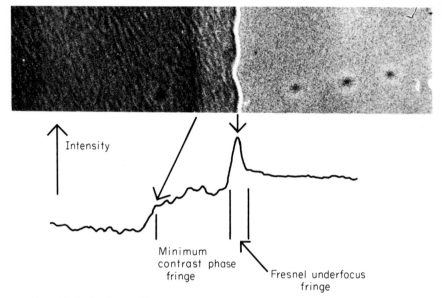

FIG. 14. This figure illustrates the *large scale* of the out-of-focus contrast effect. The underfocused carbon film shows a contrast edge effect 0.63 μ wide. This is possibly due to the "defective Zernicke condition" caused by partial mixing of phase shifted and nonshifted beams (No. 1523) or else due to separation of dark and bright field images.

observed in defocused images (Figs. 13 and 14). However, the breadth of the contrast band and also the contrast observed by us on 880 Å latex spheres and films suggest that macrocontrast can be expected to be uniform over a width of at least 1 μ. However, further work is required to distinguish phase contrast changes from contrast changes arising from separation of the bright and dark field images.

The preliminary experiments in in-focus phase contrast electron microscopy give considerable encouragement for the development of a practical phase contrast system for examination of unstained objects, both of small dimension (single molecule) and large dimension (cell structures). Much more quantitative work with thin films and electrostatic phase plates is required, and especially a better understanding of the aberrations of the objective lens are required before the best phase contrast approach can be established.

C. Out-of-Focus Phase Contrast Electron Microscopy

It has been recognized from the earliest days of electron microscopy that phase contrast effects play an important part in image
formation (*13, 99*). It is usually stated that the main reason for
observing thin objects (<10 $\mu g/cm^2$) at focus with significant contrast is the phase contrast produced by the spherical aberration of
the objective lens. This concept appears to be supported by viewing
a standard object, e.g., ferritin molecules, in microscopes of different
focal lengths and sperical aberration coefficients. The contrast appears
to be smaller in microscopes using objectives of short focal length
(e.g., 1.8 mm); however, a thorough quantitative evaluation of in-
focus contrast, including a correct calculation of the observed in-
focus contrast, has still to be made and requires careful attention.

The contrast may also be due to separation of dark and bright field
images.

The changes in image contrast with defocus have been frequently
discussed. Most attention has been paid to microcontrast changes associated with a large range of focus change. It is a common experience
for the electron microscopist to see an increase in contrast on defocusing on either side of focus (*59*). (If the illumination is made highly
coherent by using the first condenser at maximum strength, the image
may be dark on the underfocus side and light on the overfocus side
of focus.) It is also well known that these contrast changes are
much reduced by the insertion of a 50 μ objective aperture. Most light
and microwave studies of the intensity distribution around focus
(isophote curves) involve studies of the back focal plane intensity
distribution and not the intensity distribution on either side of the
image plane (*10*). Quantitative light and microwave studies of the
intensity distribution around image focus would be useful in explain-
ing some effects observed with the electron microscope. However, the
electron microscope objective analog arrangement would have to
resemble the electron microscope in the coherence of illumination,
the very large effective aperture and in the form of its wavefront
aberration.

Lord Rayleigh (*100*) showed that the image of a linear periodic
structure repeats at defocus values given by:

$$\Delta f = \pm 2nd^2/\lambda \tag{20}$$

and for a two-dimensional array of objects:

$$\Delta f = \pm 1.5nd^2/\lambda \tag{21}$$

where d is the periodicity and n is an integer. Between these points the image shows reversed contrast. Related observations were made by Conrady (16). The repeat images were termed Fourier images by Cowley and Moodie (18–21) in an extensive theoretical and experimental series of studies. They were observed in the electron microscope by Fagot and Fert (41). These authors studied objects of 50–60 Å diameter spaced apart by 80–90 Å. At 100 kV they observed that a defocus of about 12 μ reversed the contrast, and a defocus at 26 μ gave a repeat of the image. (Only the defocus on one side of focus was studied.) The effect was only observed if the illuminating beam was sufficiently coherent. An aperture of illumination of $\alpha = 0.5 \times 10^{-4}$ was used giving a coherence length in the plane of the object of $\lambda/2\alpha = 370$ Å. This extended over four periodicities and allowed a magnification of 40,000 or more. A similar study was reported by Farrant and McLean (42) using a thin section of a ferritin crystal (60 Å iron cores spaced about 110 Å apart). At 80 kV they observed the repeat Fourier image at an overfocus of 50 μ (close to the calculated value).

The contrast changes observed in defocused images must be related to the relative phase shift between the scattered wave and the nonscattered wave. This was discussed using a wave mechanical approach by von Borries and Lenz (12) and Lenz and Scheffels (83) who derived the Rayleigh relationship $\Delta f = 2nd^2/\lambda$ for a one-dimensional periodicity.

The geometrical relationship between phase shifts of paraxial rays and degree of defocus has been discussed by Scherzer (105) and Heidenreich (61). (It is often forgotten that such derivations assume that a two-dimensional ray diagram, similar to those used in light paths, can apply sufficiently accurately to the three-dimensional curved trajectories of electrons.) They obtain for the underfocus phase shift between a diffracted and nondiffracted ray:

$$\Delta\phi = -\frac{\pi}{\lambda}\Delta f\alpha^2 \tag{22}$$

Here, Δf is positive for an underfocus, and the paraxial rays are advanced (negative phase shift). The image will repeat when the phase shift is $2n\pi$, i.e., at an underfocus of Δf:

$$\Delta f = -2n\lambda/\alpha^2 \tag{23}$$

or:

$$\Delta f = -2\ nd^2/\lambda \tag{24}$$

which is the same as the Rayleigh formula [Eq. (19)].

In Eq. (24), we have defined the sign of the phase change with direction of defocus. Defocus, Δf, is considered positive for under-focus (weakened lens current, longer focal length) and negative for overfocus. The phase change is positive for underfocus representing a phase advance. A retardation (negative) of phase results from over-focus. Hence, the phase shifts are equal and opposite on each side of focus providing aberration phase shifts can be neglected. Thus, a $\pi/2$ phase shift on the underfocus side produce a *minimum* of positive contrast while a $\pi/2$ phase shift on the overfocus side produces a *maximum* of contrast.

However, the agreement is with the formula for a *linear* periodicity and not for a general two-dimensional array of scattering points. The repeat Fourier image should appear at a phase shift of $2n\pi$, the reversed contrast image at $(2n + 1)\pi$ and the enhanced contrast image at $(2n + 1)\pi/2$ phase shift (*12, 83*). Calculations predict rapid changes in contrast on slight defocusing for periodicities of 10 Å or less. Such changes in contrast are poorly documented in the literature, probably because of complex interaction of the defocus phase shift with the phase shifts due to lens aberrations and the damping effect of incoherence and dynamic phase fluctuations. It should be emphasized that determination of the defocus values (on both sides of true focus) for maximum or minimum contrast and for repeat Fourier image for a given object periodicity gives important information about the other phase shifts taking part in image formation (phase shift for scattering, spherical aberration, astigmatism, AC ripple, etc.). The effect of these additional phase shifts is to shift the defocus scale of the periodic changes in image contrast in one direction or the other.

As already mentioned, the expressions have been partially verified for large spacings (*41, 42*). Here the defocus phase shift is dominant, and aberration and focal length fluctuation phase shifts can be ignored.

It is important to note that detail observed in out-of-focus images may not relate very directly to the real structure. The detail present in an in-focus image depends on the amplitude distribution of the scattered radiation. Defocusing may minimize the contrast of the principal spacing and may maximize another periodicity which is barely visible in the in-focus image. Essentially, defocusing the objective introduces a variable narrow band pass filter which emphasizes the periodicity producing a reflection $\pi/2$ out of phase with the non-diffracted beam. Such optical effects obviously raise serious problems in interpreting fine detail of biological specimens where the micrographs are slightly out of focus. In some studies observed granularity has been described as real structure while in others it was considered to be an optical artifact due to "phase noise." The phase reinforced granularity is readily distinguished since its spacing or periodicity increases with increasing defocus, whereas a real periodicity does not.

Lenz (*12, 83*) explained the granularity appearing on defocus of carbon films entirely in terms of the defocus phase shift that introduces a $\pi/2$ phase shift:

$$\Delta f = \pm(2n + 1)\, d^2/2\lambda \qquad (25)$$

However, this approach ignores the phase shift due to spherical aberration which must be considered for periodicities $<10\,\text{Å}$, as well as other types of phase shift. Scherzer (*105*) summed the defocus and spherical aberration phase shift:

$$\phi = -\frac{\pi}{2\lambda}\,(C_o\alpha^4 - 2\Delta f\alpha^2) \qquad (26)$$

where C_o is the spherical aberration constant, and α the scatter angle. Here the spherical aberration term is integrated from zero to α. This is not justified in general, since the angular distribution of scattered intensity must also be included, and this depends on the structure under consideration. The "single ray" form of the expression is:

$$\phi = -\frac{2\pi}{\lambda}\,(C_o\alpha^4 - \tfrac{1}{2}\Delta f\alpha^2) \qquad (27)$$

Thon (*112*) claims experimentally to have disproved the Rayleigh-Lenz formula [Eq. (22)] and to have verified the Scherzer formula of Eq. (25) for the granularity appearing in defocused images of carbon films. The average spacing of light and dark image points was determined from optical diffraction patterns of the electron micrographs. However, the scattering of points of less than 10 Å spacing leaves much to be desired in this experimental test of the two defocus expressions.

The contrast changes associated with *nonperiodic* objects are much less well know. Von Albert *et al.* (*2*) studied the phase contrast of nickel particles with defocus both in the light microscope and the electron microscope. A definite sequence of contrast changes was observed in spite of the random arrangement of the particles. Johnson (*69*) applied the Rayleigh-Lentz formula to a study of the defocused images of single microfibrils, as well as linear arrays of microfibrils, and found the observed defocus enhancement to agree with the calculated values.

Recent work by Johnson and Parsons (*70*) on the contrast changes of single particles (880 Å polystyrene latex spheres) shows anomalous behavior that cannot be fitted simply to either the Rayleigh-Lenz or the Scherzer type expression. In Fig 13 a minimum of contrast occurs at an underfocus of 6 μ. On the basis of the Rayleigh-Lenz expression this implies that a linear periodicity in the particle of 68 Å or a two-dimensional periodicity of 76 Å is being emphasized by phase contrast. Since the spheres are randomly orientated and have no well defined crystal structure, corresponding three-dimensional periodicity expressions are also required. It should be mentioned that the spheres examined were sufficiently isolated from one another so that no two-dimensional periodicity effect from other spheres could be considered. In Fig. 13 the coherence was typical of medium to high resolution electron microscopy ("13 steps" on condenser I of the Elmiskop Ia). The general form of the contrast curve which increases on both sides of focus cannot be quantitatively explained by either the Rayleigh-Lenz or Scherzer expressions. The effect of increased coherence was to cause the contrast to be reversed on either side of focus as is to be expected from the Rayleigh-Lenz expression. The minimum at $\Delta f = +6$ μ cannot be explained by the Scherzer equation. The minimum implies that

$$\phi = \pi/2 = -\frac{2\pi}{\lambda}\,(C_o\alpha^4 - \tfrac{1}{2}\Delta f\alpha^2) \tag{28}$$

$$\alpha^2 = \frac{\Delta f \pm (\Delta f^2 - 4C_o\lambda)^{\frac{1}{2}}}{4C_o} \tag{29}$$

$$\alpha^2 \simeq (\Delta f \pm \Delta f)/4C_o \simeq 0 \text{ or } \Delta f/2C_o \tag{30}$$

$$\alpha \simeq \pm(\Delta f/2C_o)^{\frac{1}{2}} \tag{31}$$

$$d \simeq \lambda(2C_o/\Delta f)^{\frac{1}{2}} \tag{32}$$

This gives ($\lambda = 0.038$ Å, $C_o = 0.2$ cm, $\Delta f = 6 \times 10^{-4}$ cm) a spacing of about 1 Å, which suggests, within the accuracy of the calculation, that the wide angle nearest neighbor scatter (4–1 Å) may be responsible for the observed minimum. It remains unexplained that the macrocontrast of the 880 Å spheres is affected by the wide angle spectra phase shifts. In addition, aberration of the objective places phase shifts of hundreds of wavelengths on the spatial frequencies of about 1 Å. Clearly a more detailed study, particularly of the effects of spherical aberration in separating bright and dark field images, is required to study these effects.

The Scherzer expression has several other uncertainties that makes its quantitative application dubious. A major aim of a quantitative study of image formation in the electron microscope would be the derivation and experimental justification of a more complete and accurate expression relating the phase shifts of aberration and defocus. Notable among the deficiencies of the Scherzer expression is the absence of a term to describe the possible phase shift of elastic electron scattering. Eisenhandler and Siegel (37, 38) assumed that the phase change of elastic electron scattering is similar to that of electromagnetic radiation in that a positive phase shift or retardation of $\pi/2$ occurs on scattering. Hence, the Scherzer ("single ray") expression becomes:

$$\phi = \frac{\pi}{2} - \frac{2\pi}{\lambda}\,(C_o\alpha^4 - \tfrac{1}{2}\Delta f\alpha^2) \tag{33}$$

Recently, Heidenreich (64) has included other phase shifts (astigmatism and AC ripple) into the general expression.

It is obvious that defocusing represents the simplest way to achieve phase contrast electron microscopy. For objects of about 10 Å or less, slight changes in focus produce striking changes in contrast. However, several difficulties are encountered. The defocused image will show the granularity of "phase noise" (25) in addition to the fixed periodicity structure of the object. Defocusing may also involve a significant loss of resolution. The method appears unsuitable if it is wished to enhance the contrast of larger objects because of the large defocus required. These problems would be avoided by an in-focus phase contrast system.

The out-of-focus phase contrast approach has yet to be evaluated as a means of contrast enhancement. The Rayleigh-Lenz expression [Eqs. (20) and (21)] show that the required defocus will be proportional to the periodicity squared for a grating or lattice structure. It should be emphasized that the Rayleigh-Lenz expression applies only to a periodic structure. The out-of-focus contrast changes for single particles need reevaluation. The Rayleigh-Lenz equation indicates that high resolution can only be obtained for small periodicities where the required defocus value is small. Clearly in visualizing individual atoms the resolution loss in defocus must be balanced against the gain in contrast. To evaluate this requires a better knowledge of the scattering and lens aberration phase shifts than presently available. Calculations of Eisenhandler and Siegel (37, 38) indicated that sufficient contrast (5–10%) could be obtained for atoms with $z > 10$ with resolutions of 2.5 Å at 100 kV (or 1.5 Å at 750 kV). These authors later proposed to reduce the spherical aberration resolution limit by use of a zone plate (82) in the back focal plane of the objective (37, 38, 108).

Recently, there has been considerable application of out-of-focus phase contrast electron microscopy to periodic objects and single molecules. Komoda (79) studied defocused images of uranium oxide, molybdenum oxide and phorphyllite crystals. Heidenreich (63) examined partly graphitized carbon black by out-of-focus phase contrast using axial illumination and without using an objective aperture; he was able to resolve both the 3.4 Å and 1.7 Å spacings. It was suggested that the 3.4 Å spacing provides a useful magnification calibration, astigmatism and asymmetry check, and a convenient resolution test object. Heidenreich (63) has attempted to distinguish atomic and molecular structure from the background of "phase noise." Units of three hexagonal unit cells of graphite were visualized. In addition, the 12–15 Å α-helix of poly-γ-benzyl L-glutamate was also resolved against a background of "phase noise" granularity.

The clearest indication for use of out-of-focus phase contrast would appear to be for the study of thin crystal layers. The use of the technique for the study of amorphous, nonperiodic objects requires further study. Direct interpretation of out-of-focus images of non-periodic objects is made difficult by the presence of "phase noise" granularity with a spacing that is dependent on the degree of de-focus. It should be emphasized that the Rayleigh-Lenz expression for defocus phase shift and possibly also the Scherzer expression only apply to periodic objects. Reimer (*102*) applied the Born approxima-tion to consider the phase contrast of various arrangements of clusters of atoms as a result of the spherical aberration and defocus phase shift. Attempts were made to correlate these calculations with the observed contrast of thin films. However, the jump from calcula-tion of contrast of single atoms or small clusters of atoms to calcula-tion of contrast of larger structures of biological interest appears a formidable but not insurmountable barrier.

V. Conclusions

Biological electron microscopy has developed very rapidly in the past as a result of the discovery that relatively simple modifications of the instrument and of the specimen preparation techniques gave different types of preparation with improved imaging. Now that the need exists to examine biological material at the molecular and atomic level, more serious consideration must be given to obtaining a quantitative understanding of electron imaging.

Improving the microscope to give 1 Å resolution now appears quite feasible. However, it is no longer certain that spherical aberration alone has to be considered. Coherence, chromatic aberration, and supply instabilities may be of equal importance. It is now clear that a quantitative understanding of electron imaging depends on a more detailed knowledge of the static and dynamic phase shifts of the ob-jective lens. Detailed magnetic field mapping and phase plate experi-ments should be able to provide this information. However, it is also necessary to know the phase shifts occurring during the scattering process.

It appears likely that high voltage (1 MeV) electron microscopy will be required for biological electron microscopy. Electron diffraction experiments at conventional voltages (40–100 kV) show that ordered organic objects are disordered in seconds, whereas at high voltage such objects remain intact for longer times. This is due to the reduced inelastic scatter at high voltage. High voltage electron microscopy also makes it easier to construct water trapping specimen microchambers that will allow electron diffraction and electron microscopy of crystalline materials and cell structures in the hydrated state.

Both at conventional and high voltages, improved strioscopic dark field methods promise to be very useful for obtaining high contrast on biological objects. Preliminary work suggests that in-focus phase contrast electron microscopy, using thin film or electrostatic phase plates, is a practical possibility and some of the complications of the out-of-focus phase contrast technique can be avoided.

Finally, it appears that providing sufficient physical and theoretical work is given to the present problems of biological electron microscopy, there is a bright future ahead for the second phase of electron microscopy—the molecular and atomic resolution phase. This phase of electron microscope development promises discoveries in medicine and biology of equal or greater importance to those made in the first phase of the development of the electron microscope technique.

Because of the technical and theoretical difficulties that lie ahead, it is suggested that some type of National Electron Optics Center should be set up to help guide this work.

ACKNOWLEDGMENTS

This work was supported by Grants GB-5535 and GB-7130 from the National Science Foundation, and Grant E-457 from the American Cancer Society.

I am grateful for collaborative work and discussions with my graduate students, H. M. Johnson and P. Engler.

REFERENCES

1. Agar, A. W., Revell, R. S. M., and Scott, R. A. *Proc. Conf. Electron Microscopy, Delft, 1949* p. 52. Nijhoff, The Hague, 1949.
2. Albert, L. von, Schneider, R., and Fischer, H. Elektronenmikroskopische Sichtbarmachung von $\leq 10\,\text{Å}$ grossen Fremdstoffeinschlussen in elektrolytisch abgescheidenen Nickelschichten mittels Phasenkontrast durch Defokussieren. *Z. Naturforsch.* **19a**, 1120 (1964).
3. Bahr, G. F., Zeitler, E. H., and Kobayashi, K. High voltage electron microscopy. *J. Appl. Phys.* **37**, 2900 (1966).

4. Barer, R. Phase contrast and interference microscopy in cytology. *Phys. Tech. Biol. Res.* **3**, Part A, 1 (1966).

5. Becker, H., and Wallraff, A. Spherical aberration of magnetic lenses. *Arch. Elektrotech.* **32**, 664 (1938).

6. Bennett, A. H., Osterberg, H., Jupnik, H., and Richards, O. W. "Phase Microscopy—Principles and Applications." Wiley, New York, 1951.

7. Bethe, H. Theory of penetration of fast particles through matter. *Ann. Physik* [5] **5**, 325 (1930).

8. Boersch, H. von. Uber das primäre and sekundäre Bild im Elektronenmikroskop. I. Eingriffe in das Beugungsbild und ihr Einfluss auf die Abbildung. *Ann. Physik* [5] **26**, 631 (1936).

9. Boersch, H. von. Uber das primäre and sekudäre Bild im Elektronenmikroskop. II. Strukturuntersuchung mittels Elektronenbeugung. *Ann. Physik* [5] **27**, 75 (1936).

10. Born, M., and Wolf, E. "Principles of Optics." Pergamon Press, Oxford, 1965.

11. Borries, B. von. Electron scattering and image formation in the electron microscope. *Z. Naturforsch.* **4a**, 51 (1949).

12. Borries, B. von, and Lenz, F. Uber die Entstehung des Kontrastes im elektronenmikroskopischen Bild. *Proc. Reg. Conf. (Eur.) Electron Microscopy, Stockholm, 1956* p. 60. Academic Press, New York, 1957.

13. Bremmer, H. On a phase contrast theory of electronoptical image formation. *Natl. Bur. Std. (U.S.) Circ.* **527**, 145 (1954).

14. Burton, A. J., and Ledbetter, M. C. Electron microscopy of bacteriophage ØR: A comparison of negatively stained and untreated particles. *J. Mol. Biol.* **31**, 143 (1968).

15. Chamot, E. M., and Mason, C. W. "Handbook of Chemical Microscopy" 2nd ed., Vol. I, p. 84. Wiley, New York, 1938.

16. Conrady, A. E. Theories of microscopical vision. A vindication of the Abbe theory. *J. Roy. Microscop. Soc.* 610 (1904).

17. Cosslett, V. E. High voltage electron microscopy. *Sci. Progr. (London)* **55**, 15 (1967).

18. Cowley, J. M., and Moodie, A. F. Fourier images. I. The point source. *Proc. Phys. Soc.* **B70**, 486 (1957).

19. Cowley, J. M., and Moodie, A. F. Fourier images. II. The out-of-focus patterns. *Proc. Phys. Soc.* **B70**, 497 (1957).

20. Cowley, J. M., and Moodie, A. F. Fourier images. III. Finite sources. *Proc. Phys. Soc.* **B70**, 505 (1957).

21. Cowley, J. M., and Moodie, A. F. Fourier images. IV. The phase grating. *Proc. Phys. Soc.* **76**, 378 (1960).

22. Cox, H. L., and Bonham, R. A. Elastic electron scattering amplitudes for neutral atoms calculated using the partial wave method at 10, 40, 70 and 100 KV for $2 = 1$ to $2 = 54$. *J. Chem. Phys.* **47**, 2599 (1967).

23. Dietrich, I., Weyl, R., and Zerbst, H. High magnetic field gradient for electron microscopy. *Cryogenics* **7**, 178 (1967).

24. Dorsten, A. C. Van. The role of acceleration voltage in image formation. *Lab. Invest.* **14**, 819 (1965).

25. Dorsten, A. C. Van, and Premsela, H. F. The nature of the near focus E. M. images of amorphous substrates and its significance for obtaining high resolution pictures. *Proc. 6th Intern. Congr. Electron Microscopy, Kyoto, Japan, 1966* Vol. I, p. 21. Maruzen Co., Ltd., Tokyo, 1967.

26. Dosse, J. Exact calculation of magnetic lenses with asymmetric field distribution with $H = H_o/(1 + \{z/a\}^2)$. *Z. Physik* **117**, 316 (1941).

27. Dowell, W. C T. Electron microscopic image of the net planes of a crystal and its contrast. *Optik* **20**, 535 (1963).

28. Duffieux, M., Strioscopie des objects minces transparentes. *Bull. Soc. Sci. Bretagne* **21**, 3 (1946).

29. Dugas, J., Durandeau, P., and Fert, C. Lentilles électronique magnétiques symétriques et dyssymétriques. *Rev. Opt.* **40**, 277 (1961).

30. Dupouy, G., and Perrier, F. Amélioration de contraste en microscopie électronique. *Proc. 6th Intern. Congr. Electron Microscopy, Kyoto, Japan, 1966* Vol. I, p. 3. Maruzen Co., Ltd., Tokyo, 1967.

31. Dupouy, G., Perrier, F. and Durrieu L. L'observation de la matière vivante au moyen d'un microscopie électronique fonctionnant sous très haute tension. *Compt. Rend.* **251**, 2836 (1960).

32. Dupouy G., Perrier, F., Trinquier, J., and Fayet, Y. Condenseur-objectif pour microscopie a très haute tension. *Compt. Rend.* **265**, 676 (1967).

33. Dupouy, G., Perrier, F., and Verdier, P. *Compt. Rend.* **B262**, 1063 (1966).

34. Dupouy, G., Perrier, F., and Verdier, P. Amélioration du contraste des images d'objets amorphes minces en microscopie électronique. *J. Microscopie* **5**, 655 (1966).

35. Durandeau, P., and Fert, C. Lentilles électronique magnétiques. *Rev. Opt.* **36**, 205 (1957).

36. Dyck, R. H., and Weckler, G. P. Integrated arrays of silicon photodetectors for image sensing. *IEEE, Trans. Electron Devices* **15**, 196 (1968).

37. Eisenhandler, C. B., and Siegel, B. M. Imaging of single atoms by phase contrast. *J. Appl. Phys.* **37**, 1613 (1966).

38. Eisenhandler, C. B., and Siegel, B. M. A zone-plate aperture for enhancing resolution in phase contrast electron microscopy. *Appl. Phys. Letters* **8**, 258 (1966), see also corrections in *Appl. Phys. Letters* **9**, 217 (1966).

38a. Engler, P., and Parsons, D. F. Unpublished data (1969).

39. Faget, J., Fagot, M., Ferré, J., and Fert, C. Microscopie electronique a contraste de phase. *Proc. 5th Intern. Congr. Electron Microscopy, Philadelphia, 1962* Vol. I, Art A-7. Academic Press, New York, 1962.

40. Faget, J., Ferré, J., and Fert, C. Contraste de phase en microscopie électronique *Compt. Rend.* **251**, 526 (1960).

41. Fagot, M., and Fert, C. Contraste défocalisation en éclaïrage, cohérent, cas d'un objet périodique en microscopie électronique. *Compt. Rend.* **250**, 94 (1960).

42. Farrant, J. L., and McLean, J. D. The imaging of a protein crystal lattice.

Intern Conf. Electron Diffraction Crystal Defects, Melbourne, 1965 Abstr. IP-5.

43. Fernández-Morán, H. High resolution electron microscopy of biological specimens. *Proc. 6th Intern. Congr. Electron Microscopy, Kyoto, Japan, 1966* Vol. I, p. 13. Maruzen Co., Ltd., Tokyo, 1966.

44. Fert, C. Contraste de phase et strioscopie. *In* "Traité de microscopie électronique" (C. Magnan, ed.), Vol. I, p. 380. Hermann, Paris, 1961.

45. Fert, C., and Faget, J. Contraste de phase en microscopie electronique. *Proc. 4th Intern. Conf. Electron Microscopy, Berlin, 1958* Vol. I, p. 234. Springer, Berlin, 1960.

46. Fert, C., Faget, J., Fagot, M., and Ferré, J. Effets de diffraction et d'interférence dans les images électroniques; formation de l'image en optique électronique. *J. Phys. Soc. Japan* **17**, Suppl. B-II, 186 (1962).

47. Françon, M. "Le contraste de phase en optique et en microscopie." Editions de la Revue d'Optique Théorique et Instrumentale, Paris, 1950.

48. Glaser, W. Exact calculation of magnetic lenses with the field distribution $H = H_o/(1 + \{z/a\}^2)$. *Z. Physik* **117**, 285 (1941).

49. Glaser, W. Fundamental problems of theoretical electron optics. *Natl. Bur. Std. (U.S.), Circ.* **527**, 111 (1954).

50. Glaser, W. Elektronen und Ionenoptik. *In* "Handbuch der Physik" (S. Flügge, ed.), Vol. 33, p. 229. Springer, Berlin, 1956.

51. Glossop, A. B., and Pashley, D. W. The direct observation of anti-phase domain boundaries in ordered copper-gold (CuAu) alloy. *Proc. Roy Soc.* **A250**, 132 (1959).

52. Grivet, P. *In* "Electron Optics" (Revised by A. Septier; translated by P. W. Hawkes), p. 144. Pergamon Press, Oxford, 1965.

53. Haine, M. E., and Cosslett, V. E. "The Electron Microscope." Spon, London, 1961.

54. Hall, C. E. Dark-field electron microscopy. I. Studies of crystalline substances in dark field. *J. Appl. Phys.* **19**, 198 (1948).

55. Hall, C. E. Dark-field electron microscopy. II. Studies of colloidal carbon. *J. Appl. Phys.* **19**, 271 (1948).

56. Hanszen, K. J. Neue Erkenntnisse über Auflösung und Kontrast in elektronenmikroskopischen Bild. *Naturwissenschaften* **6**, 125 (1967).

57. Hart, R. K. *Proc. AMU (Assoc. Midwest Univ.)—ANL (Argonne Natl. Lab.) Workshop High Voltage Electron Microscopy, Argonne Natl. Lab., 1966* Rept. ANL-7275. (Available from Clearinghouse for Federal Scientific and Technical Information, National Bureau of Standards, U.S. Dept of Commerce, Springfield, Virginia.)

58. Hawkes, P. W. The dependence of the spherical aberration coefficient of an electron-objective lens on object position and magnification. *Brit. J. Appl. Phys.* **1**, 131 (1968).

59. Haydon, G. B., and Taylor, D. A. The optimal underfocus enhancement of contrast in electron microscopy. *J. Roy. Microscop. Soc.* [3] **85**, 305 (1966).

60. Heide, H. G. Electron microscope observations of specimens under controlled gas pressure. *J. Cell Biol.* **13**, 147 (1962).
61. Heidenreich, R. D. "Fundamentals of Transmission Electron Microscopy." Wiley (Interscience), New York, 1964.
62. Heidenreich, R. D. Electron phase contrast images of molecular detail. *Proc. 6th Intern. Congr. Electron Microscopy, Kyoto, Japan, 1966* Vol. I, p. 7. Maruzen Co., Ltd., Tokyo, 1967.
63. Heidenreich, R. D. Electron phase contrast images of molecular detail. *J. Electronmicroscopy (Tokyo)* **16**, 23 (1967).
64. Heidenreich, R. D. Electron phase contrast of molecular detail. *Siemens Rev.* **34**, 4 (1967).
65. Heinemann, K., and Mollenstedt, G. Elektronenmikroskopie mit langsamen Elektronen mittels Zwischenverzögerer. *Optik* **26**, 11 (1967–1968).
66. Hibi, T., and Yada, K. *J. Electronmicroscopy (Tokyo)* **13**, 94 (1964).
67. Hillier, J., and Ramberg, E. G. The magnetic electron microscope objective: Contour phenomena and the attainment of high resolving power. *J. Appl. Phys.* **18**, 48 (1947).
68. Hirsch, P. B., Howie, A., Nicholson, R. B., Pashley, D. W., and Whelan, M. J. The electron microscope. *In* "Electron Microscopy of Thin Crystals," p. 1. Plenum Press, New York, 1965.
69. Johnson, D. J. Amplitude and phase contrast in electron microscope images of molecular structures. *J. Roy. Microscop. Soc.* [3] **88**, 39 (1968).
70. Johnson, H. M., and Parsons, D. F. Four stage electron microscope for phase contrast and strioscopic microscopy. *Proc. 25th Anniv. Meeting Electron Microscopy Soc. Am., Chicago, 1967* p. **236**. Claitor's Book Store, Baton Rouge, Louisiana, 1967.
70a. Johnson, H. M., and Parsons, D. F. To be published.
71. Kanaya, K., and Kawakatsu, H. Electron phase microscope. *Proc. 4th Intern. Congr. Electron Microscopy, Berlin, 1958* Vol. I, p. 308. Springer, Berlin, 1960.
72. Kanaya, K., Kawakatsu, H., Iro, K., and Yotsumoto, H. Experiment on the electron phase microscope. *J. Appl. Phys.* **29**, 1046 (1958).
73. Kanaya, K., Kawakatsu, H., and Yotsumoto, H. On the electron phase microscope. *J. Electronmicroscopy (Tokyo)* **6**, 1 (1958).
74. Kanaya, K., Yamazaki, H., Okazaki, I., and Tanaka, K. Fundamental characteristics of electron beam exposure of photoresist and application to manufacturing diode. *Bull. Electrotech. Lab. (Tokyo)* **31**, 38 (1967).
75. Kitamura, N., Schulhof, M. P., and Siegel, B. M. Superconducting lens for electron microscopy. *Appl. Phys. Letters* **9**, 377 (1966).
76. Kobayashi, K., and Ohara, M. Voltage dependence of radiation damage to polymer specimens. *Proc. 6th Intern. Congr. Electron Microscopy, Kyoto, Japan, 1966* Vol. I, p. 579. Maruzen Co., Ltd. Tokyo, 1967.
77. Kobayashi, K., and Sakaoku, K. Irradiation changes in organic polymers at various accelerating voltages. *Lab. Invest.* **14**, 1097 (1965).

78. Komoda, T. On the resolution of the lattice imaging in the electron microscope. *Optik* **21**, 93 (1964).
79. Komoda, T. Resolution of phase contrast images in electron microscopy. *Hitachi Rev.* **49**, 43 (1967).
80. Kunath, W., and Riecke, W. D. Zur Offnungsfehlerkoeffizienten Magnetischer Objektivlinsen. *Optik* **23**, 222 (1965–1966).
81. Laberrique, A., and Severin, C. Catacteristiques optiques de lentilles magnetiques sans fer a enroulements supraconducteurs en microscopie électronique. *J. Microscopie* **6**, 123 (1967).
82. Langer, R., and Hoppe, W. Die Erhohung von Auflosung und Kontrast im Electronenmikroskop mit Zonenkorrekturplatten. *Optik* **24**, 470 (1966–1967).
83. Lenz, F., and Scheffels, W. Das Zusammenwirken von Phasen und Amplitudenkontrast in der elektronenmikroskopischen Abbidlung. *Z. Naturforsch.* **13a**, 226 (1958).
84. Locquin, M. *Proc. 3rd Intern. Conf. Electron Microscopy, London, 1954* p. 285. Roy. Microscop. Soc., London, 1956.
85. Locquin, M. Contraste de phase et contraste interchromatique étude comparée des méthodes. *Proc. 1st Reg. Conf. (Asia Oceana) Electron Microscopy, Tokyo, 1956* p. 163. Electrotech Lab., Tokyo, 1957.
86. Locquin, M. Contraste de phase et contraste interchromatique. Etude comparée des méthodes. *Proc. Reg. Conf. (Eur.) Electron Microscopy, Stockholm, 1956* p. 78. Academic Press, New York, 1957.
87. Luckiesh, M., and Moss, F. K. Light vision and seeing. *Med. Phys.* **1**, 672 (1944).
88. Mahl, H., and Recknagel, A. Uber den Offnungsfehler von elektrostatischen Elektronenlinsen. *Z. Physik* **122**, 660 (1944).
89. Martin, L. C. "The Theory of the Microscope." American Elsevier, New York, 1966.
90. Moliere, G. Theory of scattering of fast charged particles. I. Single scattering by a screened coulomb field. *Z. Naturforsch.* **2a**, 133 (1947).
91. Moretz, R. C., Johnson, H. M., and Parsons, D. F. Thickness estimation of carbon films by electron microscopy of transverse sections and optical density measurements. *J. Appl. Phys.* **39**, 5421 (1968).
92. Morrow, D. F., and Horner, J. A. Electron diffraction studies of polymer single crystals: Use of the image intensifier. *RCA Sci. Instr. News* **11**, 3 (1966).
93. O'Neill, E. L. "Introduction to Statistical Optics." Addison-Wesley, Reading, Massachusetts, 1963.
94. Parsons, D. F. Electron diffraction of helical forms of polyribonucleotides and polyamino acids. *Proc. 6th Intern. Congr. Electron Microscopy, Kyoto, Japan, 1966* Vol. II, p. 121. Maruzen Co., Ltd., Tokyo, 1967.
95. Parsons, D. F. The examination of mineral deposits in pathological tissues by electron diffraction. *Intern. Rev. Exptl. Pathol.* **6**, 1 (1968).
96. Parsons, D. F., and Martius, U. Determination of the alpha-helix con-

figuration of poly-γ-benzyl L-glutamate by electron diffraction. *J. Mol. Biol.* **10**, 530 (1964).

96a. Parsons, D. F., Hausner, G., and Moretz, R. C. Unpublished data (1968).
97. Parsons, J. R., and Hoelke, C. W. Observation of crystal lattice planes in a high vacuum Siemens Elmiskop I. *Proc. 6th Intern. Congr. Electron Microscopy, Kyoto, Japan, 1966* Vol. I, p. 37. Maruzen Co., Ltd., Tokyo, 1967.
98. Pashley, D. W., and Presland, A. E. B. The observation of anti-phase boundaries during the transition from Cu-AuI to Cu-AuII. *J. Inst. Metals* **87**, 419 (1959).
99. Ramberg, E. G. Phase contrast in electron microscope images. *J. Appl. Phys.* **20**, 441 (1949).
100. Rayleigh, Lord. On copying diffraction gratings and on some phenomena connected therewith. *Phil. Mag.* [5] **11**, 196 (1881).
101. Rebsch, R. The theoretical resolving power of the electron microscope. *Ann. Physik* [5] **31**, 551 (1938).
102. Reimer, L., Elektronenoptischer Phasenkontrast. I. Ansatz fur ene quantitative theorie. *Z. Naturforsch.* **21a**, 1489 (1966).
103. Reinhold, G. Highly stabilized accelerators for electron microscopy at high voltages. *IEEE, Trans. Nucl. Sci.* **14**, No. 3 (1967).
104. Ruska, E. Magnetic objective for the electron microscope. *Z. Physik* **89**, 90 (1934).
105. Scherzer, O. The theoretical limit of the electron microscope. *J. Appl. Phys.* **20**, 20 (1949).
106. Seifriz, W. The Spierer lens and what it reveals in cellulose and protoplasm. *J. Phys. Chem.* **34**, 118 (1931).
107. Septier, A. The struggle to overcome spherical aberration in electron optics. *In* "Advances in Optical and Electron Microscopy" (V. E. Cosslett and R. Barer, eds.), Vol. 1, p. 204. Academic Press, New York, 1966.
108. Siegel, B. M., Eisenhandler, C. B., and Coan, M. G. Ultimate resolution by phase contrast imaging of molecular objects. *Proc. 6th Intern. Congr. Electron Microscopy, Kyoto, Japan, 1966* Vol. I, p. 41. Maruzen Co., Ltd. Tokyo, 1967.
109. Simpkins, J. E. Microminiature Hall probes for use in liquid helium. *Rev. Sci. Instr.* **39**, 570 (1968).
110. Stoianova, I. G., and Mikhailovskii, G. A. Method and apparatus for studying wet objects in the electron microscope. *Biophysics (USSR) (English Transl.)* **4**, 116 (1959).
111. Taoka, T., Fujsta, H., Kanaya, K., Iwanaga, M., and Iwasa, N. Functional features of a 500 KV electron microscope. *J. Sci. Instr.* **44**, 747 (1967).
112. Thon, F. On the defocusing dependence of phase contrast in electron microscopical images. *Siemens Rev.* **34**, 13 (1967).
113. Thorvert, J. Diffusion in solutions. *Ann. Phys. (Paris)* [9] **2**, 369 (1915).
114. Toepler, A. Ueber die Methode der Schlierenbeobachtung ab mikroskopisches Hulfsmittel nebst Bemerkungen zur Theorie der Schiefer Belenchtung. *Ann. Physik* [2] **127**, 556 (1866).

115. Valentine, R. C. The response of photographic emulsions to electrons. *In* "Advances in Optical and Electron Microscopy" (V. E. Cosslett and R. Barer, eds.), Vol. 1, p. 180. Academic Press, New York, 1966.
116. Vingsbo, O. Dark field images at 1 million volt electron microscopy. *J. Microscopie* **6**, 249 (1967).
117. Watanabe, M., Shinagawa, H., and Shinota, K. Observation of metal crystal lattice image. *Proc. 6th Intern. Congr. Electron Microscopy, Kyoto, Japan, 1966* Vol. I, p. 33. Maruzen Co., Ltd., Tokyo, 1967.
118. Wilska, A. P. Expectations and limitations of low voltage electron microscopy. *Lab. Invest.* **14**, 825 (1965).
119. Yada, K., and Hibi, T. The contrast of electron images in the case of using pointed cathodes. *Proc. 6th Intern. Congr. Electron Microscopy, Kyoto, Japan, 1966* Vol. I, p. 25. Maruzen Co., Ltd., Tokyo, 1967.
120. Yada, K., and Hibi, T. Fine lattice fringes resolved by the bright and dark field axial illuminations. *Japan. J. Appl. Phys.* **6**, 1007 (1967).
121. Yada, K., and Hibi, T. Lattice images of the crystals consisting of light elements resolved by axial illumination. *Japan. J. Appl. Phys.* **7**, 178 (1968).
122. Zeitler, E. Contrast of single atoms in an aberration free electron microscope. *Proc. 6th Intern. Congr. Electron Microscopy, Kyoto, Japan, 1966* Vol. I, p. 43. Maruzen Co., Ltd., Tokyo, 1967.
123. Zernicke, F. Phase contrast, a new method for the microscopic observation of transparent objects. Part I. *Physica* **9**, 686 (1942).
124. Zernicke, F. Phase contrast, a new method for the microscopic observation of transparent objects. Part II. *Physica* **9**, 974 (1942).

Chapter II

CHEMICAL EFFECTS OF FIXATION
ON BIOLOGICAL SPECIMENS

JELLE C. RIEMERSMA

I. Introduction

The main purpose of fixation in electron microscopy is to preserve cellular organization in the course of specimen preparation. Inevitably, fixation, dehydration, and embedding result in a loss of certain cellular substances and the transformation of others, evident, for instance, from a diminished enzymic activity. The density distribution visualized on an electron micrograph does not exactly correspond to the localization of the cell components initially present. The degree to which success is achieved in the preservation of ultra-

69

structure is largely judged from the morphology observed, though also some extraneous considerations may be taken into account; for instance, in block fixation there is a greater chance of artifacts than in a rapid perfusion fixation technique. The basic question to what extent an electron micrograph represents "reality" may also be approached by means of studies comparing the image obtained by different fixation techniques. Furthermore, it would be desirable to have a comprehensive knowledge of the chemical effects of fixation, dehydration, and embedding on biological specimens.

In the following pages some of these chemical effects, namely those which occur with osmium tetroxide and permanganate fixatives, will be discussed in some detail. The equally important aldehyde fixatives have been adequately treated by other authors (76) and, therefore, have not been included. As regards dehydration and embedding, and technical prescriptions regarding specimen preparation in general, the reader is referred to a number of recent treatments (17, 18, 56, 65, 84, 90, 112).

II. Osmium Tetroxide Fixation

A. Properties of Osmium Tetroxide

A heuristic classification by Zeiger (129) distinguishes two main types of fixation reagents: (1) protein coagulants: ethanol, acetic acid, picric acid, mercuric chloride, aldehydes; and (2) lipid stabilizers: osmium tetroxide, potassium permanganate, potassium dichromate, aldehydes.

Some overlap between these categories is evident from Zeiger's listing of aldehydes under both headings. Not only aldehydes but osmium tetroxide as well may have multiple effects, contributing to protein and to lipid stabilization (45, 76, 87, 110).

Despite the fact that, in Stoeckenius' words, "our picture of the cell is still essentially a picture of the osmium tetroxide fixed cell" (117), there remains considerable uncertainty about the way OsO_4 interacts with the various classes of cellular compounds. Generally a reduction of octovalent osmium to lower oxidation states take place, but this reduction may follow different pathways, each comprising several steps (*11, 24, 41, 101*). Moreover, osmium readily forms complexes with numerous types of ligands. The outcome of a reaction with osmium tetroxide depends not only on the reaction partner but also on the nature of the solvent medium and on pH and temperature.

Osmium tetroxide at room temperature consists of yellow crystals with a melting point of 40.6°C. At 25°C the solubility in water is 7.24% (*112*). For fixation purposes 1 or 2% solutions are employed. Afzelius (*3*) has proposed the use of 40% OsO_4 in carbon tetrachloride or of molten osmium tetroxide, but the high toxicity of OsO_4 vapor makes such procedures too hazardous for common use.

The osmium tetroxide molecule has a tetrahedral structure. It contains a vacant orbital and readily adds compounds having lone electron pairs (*41*). A complex formed with one ammonia molecule is ultimately transformed to osmiamic acid (*42, 43*), as in the following equation.

$$O = \overset{\overset{\displaystyle O}{\|}}{\underset{\underset{\displaystyle O}{\|}}{Os}} = O \;+\; NH_3 \longrightarrow O = \overset{\overset{\displaystyle O}{\|}}{\underset{\underset{\displaystyle NH_3}{\diagdown}}{Os}} = O \longrightarrow O = \overset{\overset{\displaystyle O}{\|}}{\underset{\underset{\displaystyle NH_2}{\diagdown}}{Os}} - OH$$

$$H_2O \;+\; O = \overset{\overset{\displaystyle O}{\|}}{\underset{\underset{\displaystyle O}{\|}}{Os}} = N^{\ominus}\, H^{\oplus}$$

(1)

With one or more water molecules the extremely weak perosmic acid $(H_{2n}OsO_{4+n}; K_1 = 10^{-10})$ is formed; at neutral pH virtually no anions of this acid are present (*88, 106*).

Reduction of OsO_4 by potassium cyanide leads to cyanide complex of hexavalent osmium which has been characterized as a 6-coordination compound (*72*). The formula of this compound is:

$$\begin{array}{c} K^{\oplus}\; {}^{\ominus}O \diagdown \overset{\displaystyle CN}{\underset{\displaystyle CN}{\overset{|}{\underset{|}{Os}}}} \diagup CN \\ K^{\oplus}\; {}^{\ominus}O \diagup \diagdown CN \end{array}$$

Seligman and co-workers (*47, 109*) have used the complex-forming propensity of osmium for staining purposes, binding ligands to an osmate ester in aqueous medium. By means of a "sandwich" technique, the contrast produced by lipid components of the cell may be increased. A useful ligand is provided by the compound thiocarbohydrazide ($H_2N \cdot NH \cdot CS \cdot NH \cdot NH_2$). One end of the thiocarbohydrazide molecule attaches to the lipid-bound osmium; a second exposure to OsO_4 leads to osmium binding at the other end of the molecule (*109*).

B. Influence of Fixation Conditions

Osmium tetroxide penetrates rather slowly under customary conditions of tissue fixation; Zetterqvist (*130*) estimated a rate of penetration of 40 μ in 5 minutes (cf. *8, 10, 12, 44*). An adequate penetration velocity is required to prevent extensive postmortem changes which become important after 5 minutes (*54*). Added potassium dichromate accelerates penetration by OsO_4 (*125*).

Tissue blocks of 0.5–1 mm size should be cut and be immersed in the fixative as soon as possible after sacrificing the animal. Only the superficial layer, but not the core of a tissue block, is "well" fixed by the slowly penetrating osmium tetroxide (*14*); in one instance a well-preserved zone 40 μ thick was found at the surface (*93*). Excellent preservation of ultrastructure has been achieved by perfusion fixation, or by dripping fixative on an organ surface (kidney) *in vivo* (*71*).

An important variable is the time of fixation. Low (*66*) found progressive changes in lung tissue fixed for various periods in buffered 1% OsO_4. Optimal visualization of tissue components was obtained after a brief fixation period (15 minutes). More prolonged exposure to the fixing fluid caused a gradual removal of cytoplasmic material and disintegration of tissue fine structure (cf. *87*). Usually a fixation period of 1–2 hours is adequate, while with microorganisms the time may be shortened (*7, 76, 112*). The influence of temperature has not been extensively studied; it is accepted practice to fix at 0°–4°C.

After Palade's observation (*80, 81*) that osmium tetroxide reacts with tissue components in such a way that acidification occurs, fixation has generally been carried out in buffer media. With regard to particular tissues, for instance renal tubular epithelium (*93*) and intestinal epithelium (*130*), a controlled pH is very important. In other cases there may be more latitude, and certain cases of tissue damage, initially attributed to adverse pH, are more probably artifacts of methacrylate embedding (*112, 121*).

At present, in the opinion of Wood and Luft (*128*), "the electron microscopist has less certainty about the concept of a simple 'best' fixation but more flexibility in choice of fixative for specific purposes." An early fixation medium was the veronal–acetate buffer of pH 7.3–7.5 introduced by Palade (*80*) and modified by Rhodin (*93*) and Zetterqvist (*130*). Phosphate buffers are more effective in maintaining a constant pH and are superior to veronal–acetate mixtures in preserving cellular ultrastructure (*76*). Collidine buffers are also widely used; it has been claimed that the fixed material is more easily sectioned than material fixed in other buffer media (*107, 128*). The chrome–osmium fixative recommended by Dalton (*27*) should also be mentioned. For the composition of the main fixative mixtures several recent books may be consulted (*56, 84, 90, 112*).

C. Water Movements

When tissue slices are brought into an osmium tetroxide solution there is a characteristic rapid volume increase. Bahr (*8–10*, cf. *125*) studied volume changes in fixation and found that swelling occurred even in hypertonic media. Tissue specimens after immersion in an OsO_4 solution showed an increased weight and specific gravity, which indicates that not only water movements but also an uptake of osmium in bound form occurred. The period of rapid swelling coincided with the period of osmium uptake as revealed by chemical analysis (*9*).

In one series of experiments, slices of liver tissue (150–200 mg) were fixed in 1% OsO_4, in balanced Tyrode solution at pH 7.2 during 24 hours. Half-maximal swelling was observed already after 15 minutes. Maximal swelling, after 4 hours, corresponded to a 30% volume increase of the tissue. Somewhat different findings were obtained with formaldehyde fixatives; here, swelling occurred more slowly. As with the osmium fixatives, increasing the ion strength of the formaldehyde solutions gave a diminished swelling effect, while swelling was counteracted strongly by adding to the fixation medium dextran or other macromolecular substances (9, 125).

The extent of volume increase during fixation, whether osmium tetroxide or aldehyde fixatives are used, undoubtedly depends on tonicity. Only in fixation media of controlled tonicity can an optimal preservation of ultrastructure be obtained (70, 93, 129, 130). Exactly isotonic conditions, however, are hardly definable during fixation because: (1) The permeability of cell membranes changes as a result of the actions of osmium tetroxide; and (2) new anionic groups are formed in addition to the ones already present.

After osmium fixation cell membranes appear to become completely permeable to small ions. In a study of such permeability changes in *Limnaea* eggs, Elbers (29) found that a Donnan equilibrium was established between eggs and medium when the eggs were brought in 1% OsO_4 in distilled water (pH 6); considerable swelling of the eggs occurred. This swelling was much reduced, to about 10%, when 2% OsO_4 was used in a salt medium isotonic with the cytoplasm. An OsO_4 solution in 142 mM NaCl, 5 mM KCl, 2 mM $CaCl_2$, 1 mM $MgCl_2$, and 1 mM NaH_2PO_4 gave optimal results.

The role of tonicity in fixation has also been demonstrated in the case of photoreceptors (97). Osmium tetroxide solution of low tonicity gave swelling of the retinal sacs and separation of membranes, while better results were obtained at higher tonicity. As another example, from prawn nerve sheath's "normal" electron micrographs were obtained only when fixation was carried out at the appropriate osmolarity (800 mOsmols) (28).

Osmium tetroxide greatly alters the permeability of plasma membranes, which for instance have shown to permit even the passage of ferritin molecules (119). Nevertheless dextran or polyvinyl pyrrolidone molecules present in the fixative solution oppose cell swelling; they probably remain outside the cells during the fixation process because of their low rate of diffusion (8, 9, 63, 70, 128). Such additives often have a beneficial effect in regard to tissue preservation, although the rate of OsO_4 penetration appears to be slowed down.

After fixation the tissue blocks are brought into the first of a series of dehydrating agents with or without previous washing. There seems to be little objection to the use of water, although usually isotonic salt solutions are used for rinsing.

The plastic embedding media used in electron microscopy can be distinguished into two main classes, depending on whether their monomers are water-soluble or water-insoluble. Methacrylate, epoxy resins (araldite, Epon), and certain polyesters (Vestopal W) belong to the second category; water-miscible embedding media are Aquon, Durcupan, and glycol-methacrylate. When water-immiscible media such as methacrylate are used, the tissue block has to be completely dehydrated before embedding. The changeover from a watery to a nonaqueous medium has to be carried out stepwise to avoid disruptive effects of surface tension. The gradient of the solvent series, e.g., ethanol–water mixtures with decreasing percentage of water, and the time schedule of dehydration are not highly critical, although dehydration should be carried out at sufficient speed to prevent excessive extraction (77). Dissolution of cellular substances can be counteracted by adding a salt, such as magnesium chloride, to the earlier solvents of the series. The use of salt-containing fixatives previous to dehydration appears to diminish the loss of cellular material also, while sucrose-containing fixatives have the opposite effect (68, 75, 76, 118).

Dehydration is accompanied by tissue shrinkage to an extent which compensates largely the swelling of tissue during fixation, whereas the embedding media usually cause no further shrinkage (77). Since the replacement of water by a relatively polar solvent can be expected to lead to a denaturation of cellular proteins and hence to ultrastructural changes, the possibilities have been explored of circumventing such solvents in tissue dehydration. In combination with appropriate embedding media, dehydration can be carried out with ethylene glycol and other polar solvents less likely to lead to protein denaturation (85, 86).

D. CHEMICAL EFFECTS OF OSMIUM TETROXIDE

Direct as well as indirect evidence points to a participation of cellular lipids in the reactions leading to osmium deposits after osmium tetroxide fixation. On electron micrographs the membranes of cells and cellular organelles are characteristically represented in cross section as two dark bands with a light band in between. Similar alternations of dark and light bands were also seen on electron micrographs of lipid–water preparations, so-called myelinic forms (*4, 21, 33, 91, 108, 115, 116*). In such lipid–water preparations, the lipid molecules are known to be present in a regular array, in platelike micelles, as evident from optical and x-ray data (*49, 78, 114*). In agreement with a widely adopted model of the cell membrane, the trilaminar appearance of cell membranes in cross section may be interpreted as a picture of a membrane containing as its central part a lipid bilayer in agreement with the Gorter-Danielli-Robertson model (*34, 35, 37, 92, 98, 99*). Such an interpretation remains, however, uncertain as long as the chemistry of osmium fixation is inadequately known (*110*).

Recent findings concerning hydrophobically bonded proteins in biological membranes do not invalidate the traditional Gorter-Danielli-Robertson bilayer model, but they do necessitate its revision as to the role of membrane proteins (*123*). Structural proteins intimately associated with lipids may well contribute to membrane image formation. In recent years the reaction between OsO_4 and various unsaturated substances has been studied in some detail. The resulting findings can be briefly summarized, with reference to possible mechanisms of fixation and staining.

At or below room temperature unsaturated lipids are far more reactive toward osmium tetroxide than are saturated lipids (*13, 21, 33, 83, 115*). A reactivity of saturated lipids has been demonstrated only at relatively high temperatures. Saturated phosphatidyl ethanolamines, for instance, are reactive at about 60°C. The reactions at high temperature may not be of the same type as the reactions with unsaturated compounds at room temperature; in the latter, primarily double bonds are involved and in the former probably the ethanolamine group (*22*).

The study of the reactions of compounds containing unsaturated carbon chains with osmium tetroxide has been hampered by the fact that usually, under preparative conditions, tarry mixtures of reaction products are formed from which it is difficult to obtain identifiable compounds. With acenaphthylene, indene, and certain other unsaturated substances, Criegee (*24–26*) obtained osmic acid monoesters and diesters as well-defined intermediates.

The localization of osmium deposits in fixed tissue may be essentially determined by the nature of the primary intermediates formed when lipids react with osmium tetroxide. A list of osmic acid derivatives that may be considered in this connection is given in Table I.

In a watery solution containing 30% *tert*-butyl alcohol, a medium relatively inert toward OsO_4, various unsaturated compounds react with OsO_4 by forming a compound characterized by a single pronounced absorption maximum (470 mμ). In 30% *tert*-butyl alcohol-water, long-chain lipid molecules probably exist as micellar aggregates of restricted size (*30*). On the basis of the absorption spectrum of the reaction products, and using the method of continuous variations (*55, 57*), the mixing ratio at maximal absorption was found to be 1 mole OsO_4 per double bond (*95*). The stoichiometry of the primary reaction between an unsaturated compound and OsO_4 indicates monoester formation.

A single absorption maximum at 470 mμ can be observed not

TABLE I
Types of Osmic Acid Derivatives

Compound	Formula
Cyclic monoester (Ia)	
Hydrated cyclic monoester (Ib)	
Cyclic diester (Os^{VI}) (II)	
Cyclic diester (Os^{IV}) (III)	
Potassium osmate (IV)	

only when unsaturated compounds in watery solution react with OsO_4, but also when dihydroxy compounds react with potassium osmate ($K_2OsO_4 \cdot 2\ H_2O$). By means of conductometric titration curves, the latter reaction has been shown to lead to the formation of cyclic osmic acid monoesters (95). Such monoesters are stable in watery solution; their formation by two different reactions can be described by Eqs. (2) and (3).

$$\tag{2}$$

$$\tag{3}$$

In experiments on a preparative scale, a different type of cyclic osmic ester was obtained, namely a diester (59, 61). A pure unsaturated compound, for instance methyl oleate, was reacted in the form of oily droplets with a watery OsO_4 solution. From the resulting tarry mixture of reaction products, after extraction with heptane and subsequent purification, a compound was isolated having the properties of an osmic acid diester. While the monoesters dissolve in watery media, diesters are heptane-soluble. Diesters have an absorption spectrum with two maxima at 465 and 565 mμ, respectively (Fig. 1). Thus, recent work has led to the conclusion that, depending on the reaction conditions, different types of osmic ester intermediates can be obtained (see also 60, 61).

A role of the solvent in the reaction between OsO_4 and an unsaturated compound, such as pure egg lecithin, is evident from the different absorption spectra obtained in carbon tetrachloride and 30% tert-butyl alcohol–water, respectively (Fig. 2). The 470 mμ absorbing compound is not formed in carbon tetrachloride. The 470 mμ peak was also absent when the reaction with OsO_4 took place in methanol, dioxane, or glacial acetic acid. Since it is found only in water-containing media, such as 80% acetone or 30% tert-butyl alcohol, it appears likely that this compound has a hydrated structure, such as compound Ib of Table I.

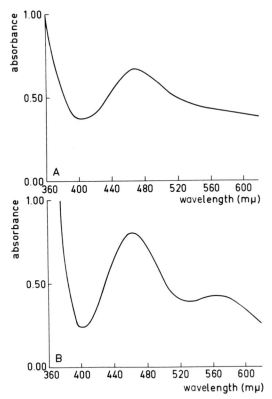

FIG. 1. Absorption spectra of OsO$_4$–methyl oleate reaction products obtained in different procedures. (A) Dissolved methyl oleate: 0.2 ml 0.02 M methyl oleate in 45% *tert*-butyl alcohol + 0.2 ml 0.02 M OsO$_4$ (H$_2$O) + 5 ml 45% *tert*-butyl alcohol; temp. = 4°C, reaction time = 17 hours. (B) Liquid methyl oleate (250 mg) + 30 ml 2% OsO$_4$; temp. = 0°C, reaction time = 3 hours. Product was extracted with heptane and purified over an SiO$_2$ column (eluate spectrum). [From Riemersma (*95*) reproduced by courtesy of Elsevier Publishing Co., Amsterdam.]

Reactions leading to hydrated osmic esters may be of the following type.

$$\begin{array}{c}
\diagdown \diagup \\
\text{C} \\
\parallel \quad + \ \text{OsO}_4 \ + \ \text{H}_2\text{O} \ \longrightarrow \\
\text{C} \\
\diagup \diagdown
\end{array}
\qquad
\begin{array}{c}
\diagdown \diagup \\
\text{C}-\text{O}\ \text{OH} \\
\quad\quad\ \ \text{Os}{=}\text{O} \\
\text{C}-\text{O}\ \text{OH} \\
\diagup \diagdown
\end{array}
\qquad (4)$$

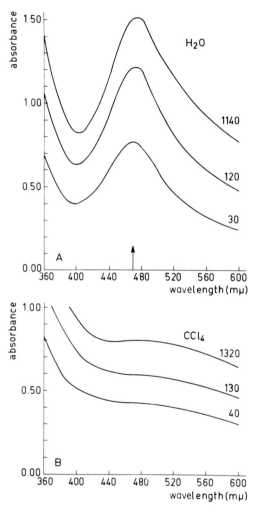

FIG. 2. Absorption spectra of the OsO_4–lecithin reaction products in two different solvents. (A) 5 ml 0.4 mM lecithin in 30% *tert*-butyl alcohol $+$ 0.15 ml 20 mM OsO_4 (H_2O). (B) 5 ml 0.4 mM lecithin (CCl_4) $+$ 0.15 ml 20 mM OsO_4 (CCl_4). 1 cm cuvettes used. Reaction times indicated in minutes; temp. $= 20°C$. [From Riemersma (*95*).]

Nonhydrated compounds, such as type Ia of Table I, are nonionizable. Hydrated compounds on the other hand, such as Ib (Table I), or the reaction product of Eq. (4) are most probably ionizing compounds. In watery media the reaction products are acidic, as is evident from the pH fall which results when OsO_4 is added to a solution of 1-decene, crotonic acid, lecithin, or methyl oleate (Fig. 3). In the absence of water, for instance in pure methanol, unsaturated substances give neither a pH fall nor a reaction product with an absorption peak at 470 mμ. It appears likely, therefore, that osmic acid monoesters of a

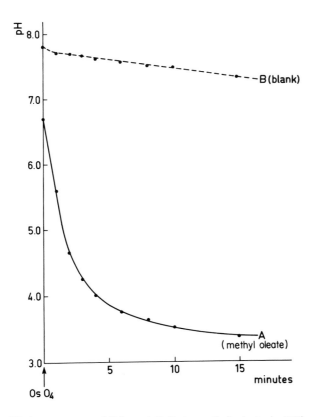

FIG. 3. pH changes upon addition of OsO_4 to methyl oleate in 75% acetone as a solvent. (A) 5 ml 0.02 M methyl oleate $+$ 15 ml solvent; addition at $t = 0$ of 5 ml 0.02 M OsO_4. (B) 20 ml solvent, at $t = 0$ of 5 ml OsO_4. Temperature $=$ 20°C. (Riemersma, unpublished experiments.)

hydrated type are formed only in such media as 75% acetone (Fig. 3) or 30% *tert*-butyl alcohol (Fig. 2). The pH decrease observed when tissue is brought into an OsO_4 fixative is probably due to the presence of unsaturated acyl chains in membranous materials.

The anions resulting from the ionization of osmic monoesters readily combine with cationic groups of amphiphilic compounds giving insoluble precipitates. Such a precipitate is formed in the experiment illustrated by Fig. 3, if instead of methyl oleate the solution contains both methyl oleate and a quaternary ammonium salt (for instance, octadecyl trimethyl ammonium chloride) and if to this solution OsO_4 is added.

To assess the stoichiometry of the primary reaction between phosphatides and OsO_4, apart from experiments in watery solution, dried lecithin spots on a glass or filter paper carrier were exposed to OsO_4 vapor. No other substances are present besides lipid and OsO_4, other than a small quantity of hydration water which is hard to remove; the carriers used are virtually inert. Drops of lecithin solution (20 μliter) were dried on cover glasses and dried to constant weight. Subsequently they were placed in closed containers with a few crystals of osmium tetroxide. The weight increase of the spots could be determined with a sensitive balance, as a function of the time of exposure to OsO_4 and subsequent treatment.

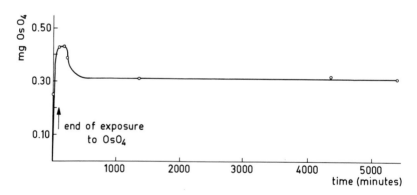

Fig. 4. Weight increase of spots of pure egg lecithin (0.66 mg per spot) on cover glasses, upon exposure to osmium tetroxide. Temperature $= 20°C$. [From Riemersma and Booij (*96*) reproduced by courtesy of Williams & Wilkins Co., Baltimore.]

When the weight increases due to OsO_4 was plotted as a function of time, the resulting curve had an "overshoot" due to the uptake of OsO_4 in a dissolved but unbound state; this quantity could be removed by standing in air (Fig. 4). The remaining weight increase, corresponding to bound OsO_4, had a linear relation to the quantity of lecithin present (Fig. 5). From the slope of this straight line a binding ratio of 1.5 moles OsO_4 per mole lecithin could be calculated. Since the fatty acid composition of the lecithin used was such that per mole, 1.5 double bonds were available, this corresponded to 1 mole OsO_4 per double bond (*96*).

In spot-test experiments evidence was obtained that exposure to osmium tetroxide changes the properties of the lecithin polar groups, particularly as regards stainability by organic dyes. The spots, applied for instance on cover glasses, are normally stainable in a bath containing 0.1% Ponceau 4 RS (Brilliant Scarlet 3 R, Color Index No. 16255), 0.4% uranyl nitrate, and 0.01 N HCl, to the extent that per mole lecithin one-third mole Ponceau 4 RS is bound. This means that lecithin, a bipolar phosphatide, reacts with uranyl ions and dye anions by forming a "tricomplex compound" on a basis of charge equivalence (*51*). Much less dye is bound by lecithin spots previously exposed to osmium tetroxide vapor. When the quantity of dye taken up per spot is plotted as a function of the time of preexposure to OsO_4, the curve of Fig. 6 is obtained (*96*). The properties of the

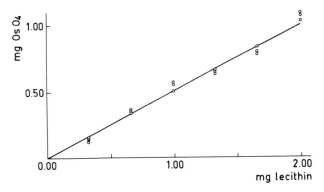

Fig. 5. Weight increase of spots of pure egg lecithin on cover glasses after exposure to OsO_4 (20°C), as a function of the lecithin quantity. [From Riemersma and Booij (*96*) reproduced by courtesy of Williams & Wilkins Co., Baltimore.]

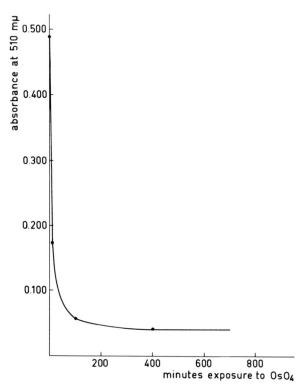

Fig. 6. Binding capacity of lecithin (0.195 mg on cover glasses) for Brilliant Scarlet 3 R as a function of the time of previous exposure to OsO₄. The spots are brought in OsO₄ and stained; subsequently, the dye is eluted. Eluate absorbance at the absorption maximum of the dye (510 mμ) is a measure of the dye quantity present. [From Riemersma and Booij (96) reproduced by courtesy of Williams & Wilkins Co., Baltimore.]

lecithin spots are changed in such a way that a large part of the potentially available choline cations no longer interact with dye anions. To interpret this phenomenon the possibility should be considered that the anionic groups of osmic acid esters combine with choline groups which are thereby no longer accessible to dye anions. Such an intramolecular salt formation could result in the observed change in dye binding capacity of lecithin spots.

While evidence regarding cyclic osmic esters has been largely derived from experiments with isolated substances, similar compounds may be formed during fixation. The situation is, however, made complex by the presence of cellular compounds other than lipids and by the action of dehydrating solvents with reducing properties. Various authors have distinguished "primary blackening" of tissue by OsO_4 from "secondary blackening" which results upon treatment with ethanol. If a nonreducing solvent such as tetrahydrofuran is used in dehydration, the secondary blackening is absent. It has been shown that cell membranes show up indistinctly, if visible at all, when tetrahydrofuran-dehydrated specimens are used in electron microscopy (*58, 108*).

From experiments with isolated compounds only tentative conclusions can be derived about the interaction between cellular lipids and OsO_4 in tissue preparation. It appears reasonably certain that lipid double bonds are an important locus of attack by osmium tetroxide, while it may also be assumed that cyclic osmic monoesters are a primary, if transient, intermediate. By interacting with cationic groups belonging to phosphatide molecules, the monoester anionic groups could move into the polar layers of lipid membranes, and thus the ultimate location of osmium deposits as visible on electron micrographs would be determined. Reduction of osmic ester by reducing substances present in cellular material, as well as by the dehydrating solvent, would lead to osmium dioxide hydrate as well as organic end products (*24, 94*). Morphological evidence also strongly suggests that osmium deposition occurs at polar groups at the interface between cellular lipids and proteins.

The reactions of substances other than lipids, such as carbohydrates, nucleic acids, and proteins, have been extensively studied—mainly by qualitative methods. Carbohydrates react slowly with osmium tetroxide, without deposition of osmium dioxide but with formation of carbonyl groups. According to Wolman (*127*), the carbonyl groups formed with glycogen, mucein, and other compounds can be demonstrated with Schiff reagent. The low electron density of glycogen in osmium-fixed tissue necessitates the use of special stains such as lead hydroxide.

Nucleic acids are relatively inert toward osmium tetroxide (*6*). A specific reactivity of thymine has been discovered in a search for base-specific markers. Thymine in a nucleotide chain reacts with OsO_4 forming a cyclic thymidine-osmate ester which by the addition of cyanide becomes stable in water (*16, 50*).

Functional groups with reducing properties as found in protein side chains determine the reactivity of proteins toward osmium tetroxide. According to Bahr such groups can be arranged in a reactivity sequence as follows:

$$-SH > C{=}C > \text{terminal} -NH_2 > -S-S- > C{=}O >$$
$$\text{terminal} -OH > \text{aromatic} -OH$$

Proteins are reactive toward OsO_4 mainly because of their tryptophan, cysteine, and histidine residues (6). According to Porter and Kallman (87), fixation by OsO_4 is the result of a partial denaturation and insolubilization of cell proteins; this view, of course, does not conflict with Bahr's conception of protein reactivity toward OsO_4 (6).

Millonig (75, 76) experimented with pure proteins, particularly albumin, which under appropriate conditions were converted by OsO_4 into gels. Gel formation was found to require fairly concentrated solutions, and did not occur, for instance, in a solution containing 5% albumin and 1 or 2% OsO_4. A gel was formed in an 8% albumin solution containing 2% OsO_4 after 3–6 hours at room temperature (76). The required protein concentrations were very high; various proteins other than albumin have been converted into gels at lower concentrations (87). In the presence of salts, gel formation is strongly promoted. With 0.8% NaCl present, 1% OsO_4 "fixes" 8% albumin, and 2% OsO_4 "fixes" 5% albumin. Salts accelerate gel formation when the reaction between OsO_4 and protein has progressed beyond a certain point; their effects may be ascribed to electrostatic shielding of charged groups. During OsO_4-fixation in a medium containing calcium ions, a considerable quantity of calcium is bound (63). Fixation in the presence of divalent ions creates a granular appearance of cell structures on electron micrographs; monovalent ions do not have such an effect (75, 76, 118). Apart from the salt added to the fixation medium, salts present in the cytoplasm may accelerate the formation of a cytoplasmic gel. A predominant cellular cation is the potassium ion, which is, however, less effective than sodium in this respect.

It appears likely, though by no means proved, that protein interlinkages contribute to the fixation effect of osmium tetroxide on cellular material. Proteins do not contribute greatly to the formation of osmium deposits (45, 75, 76). Tissue osmiophilia was not much reduced by a pretreatment with reagents eliminating protein side chains from reacting with osmium tetroxide, whereas reagents eliminating double bonds of lipids had a considerable effect (*1, 2*). Whatever the role of lipids in fixation, they must be mainly responsible for osmium staining.

As a result of the action of osmium tetroxide, the isoelectric point of proteins is lowered, indicating a disappearance of basic groups (*118*). This is a remarkable phenomenon, given the relative inertness of lysine and similar NH_2-containing amino acids toward OsO_4.

The reactions with osmium tetroxide may affect not only the normally dissolved proteins of the cytoplasm but also the structural proteins forming an integral part of cell organelles, such as mitochondria (*76, 111*). Mitochondria, from which solvent extraction had removed about 90% of the initial lipid content, still showed after osmium fixation the triple-layered membranes typical of intact mitochondria (*36*). This suggests not only that the structure of mitochondria may be preserved by substances other than lipids, but also that these substances can determine image formation after osmium treatment.

In this connection it may be mentioned that glutaraldehyde as well as osmium fixation leads to the preservation of membrane ultrastructure. Since glutaraldehyde is a protein reagent, this fixative may preserve membranes by an action on structural proteins. If following glutaraldehyde treatment, the lipids are largely removed by dehydrating solvents, membrane structure is not obliterated but remains visible in "negative contrast"; hydrophobic areas appear as gaps in the membrane structure. Again these findings indicate that membrane integrity does not require the presence of lipids. A role of glutaraldehyde in cross-linking lipid NH_2-groups, as suggested by Anderson and Roels (*5*), appears unlikely.

III. Permanganate Fixation

Potassium permanganate fixation outlines cellular membranes but extracts a large part of the cellular proteins and ribonucleic acids (*36, 76*). Like osmium tetroxide, permanganate is a staining as well as a fixation reagent (*15*). Its oxidant properties are pH-dependent, and buffered solutions should be employed to obtain controlled fixa-

tion conditions. A widely used fixation medium consists of 1.2%
potassium permanganate in an acetate–barbiturate buffer at pH 7.4
(67). Permanganate fixatives are generally used at a temperature of
0°–4°C. The nature of the cation appears to be relevant; it has been
claimed that sodium permanganate gives improved preservation of
the plasma membrane as compared to potassium permanganate (3,
126).

The similarity of pictures obtained with osmium tetroxide and
with permanganate fixatives suggests that, in some respects, the
chemical actions of these fixatives are similar. Both reagents are
oxidants, permanganate being the stronger oxidant of the two (101).
Both interact with unsaturated lipids and many other types of substances (45). The assumption that permanganate is essentially a
protein reagent (cf. 3, p. 17) is not widely shared. Although proteins,
polysaccharides, and nucleic acids can react with permanganate and
may contribute to image formation, the substances localized on
electron micrographs appear to be mainly lipids.

Potassium permanganate as well as osmium tetroxide, after reacting with pure lipids, ultimately produce lower oxides of manganese
and osmium, respectively. Oxide deposits seem to occur along the
polar strata of lipid bilayers (49). Permanganate gives an excellent
delineation of membranous structures probably because membranes
contain relatively dense aggregates of unsaturated acyl chains.

In a series of experiments with two lipid substances, oleic acid and
lecithin, their solutions were applied on filter paper; the resulting
spots were dried and immersed for varying periods in buffered
permanganate solution. The reaction conditions resembled those customary in fixation for electron microscopy (0.038 M KMnO$_4$, temperature 0°C, pH 7.4). Instead of 0.038 M permanganate (67), a 0.001 M
permanganate solution was used to reduce the background coloration
of the filter paper and to obtain a more gradual staining of the lipid
spots. The spots now appeared as dark brown areas against a negligi-
bly stained background. In reactions of this type the inorganic reaction product is manganese dioxide (74).

The circular spots (13 mm diameter) on the filter paper carriers
were left in the staining bath for the required period, washed with
distilled water, drained on filter paper, and dried in air at room
temperature for 1 hour. With a Joyce and Loebl densitometer, the

staining intensity of the spots could be measured (*64*). After a staining period of about 1 hour the densitometrically determined staining intensity reached a saturation value.

Manganese uptake was determined by chemical estimates of the
manganese contained in each spot after a given staining period. Spots
of oleic acid and of lecithin were stained 2 hours in 0.002 M KMnO$_4$
at pH 7.5 (4°C). Lengthening the staining period to 3 hours did not
increase Mn uptake. After washing in distilled water and drying in
air, the spots were cut out with scissors and eluted for 3 hours with
0.3 ml 6 N hydrochloric acid. Then 1 ml water and 0.3 ml formaldoxime
reagent were added, and immediately afterwards 1 ml 6.5 N ammonia
(*40*). The volume of solution was brought to 25 ml. A manganese–
formaldoxime complex is formed, the absorbance of which at 450 mμ
is a measure for the manganese quantity per spot. Manganese uptake
by a filter paper circle with the diameter of the spot was subtracted
as a blank.

The manganese content was determined of spots with different
lipid quantities, for oleic acid and for lecithin. In the latter
case a linear relationship was obtained between lipid quantity and
manganese uptake; in the case of oleic acid this was true only
in the range of small lipid quantities. In this initial range per mole
oleic acid, 1 mole Mn was bound and per mole lecithin (1.5 double
bonds per molecule), 3.5 moles Mn (Fig. 7). Thus, per double bond
in oleic acid, 1 mole Mn is bound; whereas in the case of lecithin, the
ratio is 2.5 moles Mn per double bond. The nature of the polar group
codetermines the course of permanganate reduction by unsaturated
compounds.

Manganese dioxide has amphoteric properties and thus can interact
with negative and positive groups or organic molecules (*74*). Manganese dioxide deposits may, therefore, be expected along the polar
faces of lipid micelles. Such a location is compatible with the observed influence of polar groups in permanganate reduction, and also
with the fact that poststaining by a uranyl salt intensifies but does
not change the membrane image given by permanganate (*79*). The
sequence of steps of permanaganate reduction includes formation of
cyclic manganic esters as intermediates, although such compounds
are not stable enough for isolation and study of their properties (*124*).
A primary step in this reduction is represented by Eq. (5):

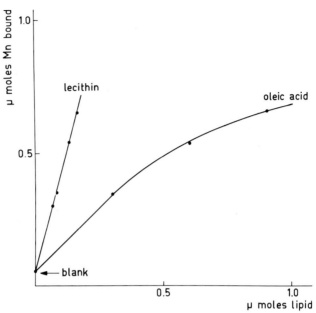

Fɪɢ. 7. Manganese present in spots of lecithin and oleic acid on filter paper (Schleicher & Schüll 2043 b Mgl) after staining in KMnO₄ solution, as a function of lipid quantity. Staining solution 0.002 M KMnO₄ in veronal–acetate buffer pH 7.5 (*67*); staining period = 2 hours, temp. = 4°C. (Riemersma, unpublished experiments.)

$$\begin{array}{c}
\diagdown\diagup \\
\mathrm{C} \\
\| \\
\mathrm{C} \\
\diagup\diagdown
\end{array}
\;+\;
\begin{array}{c}
\mathrm{O}\quad\mathrm{O} \\
\diagdown\!\diagup \\
\mathrm{Mn} \\
\diagup\diagdown \\
\mathrm{O}\quad\mathrm{O}^{\ominus}
\end{array}
\;\longrightarrow\;
\begin{array}{c}
\diagdown\diagup \\
\mathrm{C\!-\!O}\quad\mathrm{O} \\
\mid\quad\diagdown\!\diagup \\
\quad\mathrm{Mn} \\
\mid\quad\diagup\diagdown \\
\mathrm{C\!-\!O}'\;\;\mathrm{O}^{\ominus} \\
\diagup\diagdown
\end{array}
\qquad (5)$$

which resembles the formation of cyclic osmic acid esters (*48*).

Which organic endproducts are obtained after the reaction between permanganate and oxidizable organic compounds depends on the nature of the compound, pH, and temperature (*69*). Permanganate above pH 7 liberates ammonia from peptides and proteins, while the peptide linkages remain largely intact. In experiments with wool protein as test material, KMnO₄ produced considerable damage, however, in the protein fiber; in contrast, OsO₄ gave excellent preservation (*45*). Unsaturated lipids, under the customary conditions of

KMnO₄ fixation, probably yield ketohydroxy compounds and diketo compounds (*23*). At higher pH, for instance 11–12, dihydroxy compounds are formed.

Both with regard to lipids and with regard to proteins, potassium permanganate is more destructive than osmium tetroxide, but it remains a useful reagent in ultrastructure investigation. Sjöstrand, on the basis of micrographs obtained after permanganate fixation, differentiated between organelle membranes and the plasma membrane. After rapid permanganate fixation, the polar strata of mitochondrial membranes were connected by opaque cross-bridges, while in plasma membranes no such bridging was found (*111*).

Comparisons have been made between permanganate and osmium tetroxide fixation as regards ultrastructural morphology (*3*). After OsO₄ treatment of toad spinal ganglia, Rosenbluth (*100*) found a breakdown of cell invaginations; vesicles were formed which apparently enclosed extracellular material. Permanganate fixation, on the other hand, gave vesicles with extracellular material on the outside and cytoplasm enclosed. Doggenweiler (*28*) has described a similar membrane instability for prawn nerve sheaths; here the type of vesiculation appeared to depend on the osmolarity rather than on the composition of the fixative. More work is needed on the chemical aspects of permanganate fixation before such findings can be adequately interpreted (*39*).

IV. Recent Techniques of Specimen Preparation

At present in addition to the long-standing fixatives such as osmium tetroxide and potassium permanganate, a number of other fixatives have acquired equal importance. Glutaraldehyde combined with postfixation with osmium tetroxide is most widely used (*90*). The advantage of aldehyde fixation is the possibility of preserving enzyme activity to be revealed by enzyme reactions following fixation. Since the aldehydes give little increase of electron density, they have to be usually complemented by contrast-enhancing reagents. Protein cross-linking appears to be the general mode of action of aldehydes (*76, 105*). Higher aldehydes, such as glutaraldehyde and hydroxyadipaldehyde are an improvement over formaldehyde as regards enzyme preservation (90, 104, 105). In many cases glutaraldehyde gives best results, but the choice of a suitable fixative remains to be determined according to the type of cells to be fixed and the enzymes to be localized (31).

To avoid chemical changes altogether, sections can be prepared for electron microscopy from embedded frozen–dried tissue (113). The rate of cooling and other experimental variables have to be carefully controlled. Artifacts can result from ice-crystal formation and from the high salt concentrations occurring in unfrozen cytoplasm when part of it is frozen; the cell surface is also highly sensitive to freezing injury (120). Dehydration takes several days.

Next to freeze-drying, freeze-substitution is a technique which may contribute greatly to the clarification of cell anatomy. After freezing, the tissue is in contact with a dehydrant such as acetone or ethanol at −30° to −60°C for 2 weeks or a month (89). Before freezing, the tissue has to be impregnated with glycerol or dimethyl sulfoxide to prevent ice artifacts. Important variables are pretreatment, sample size, and the manner of introducing the sample into the cooling fluid; depending on the procedure employed, different morphologies are obtained (20). Addition of 1% OsO_4 appeared to have a positive effect (89). So far experimental difficulties prevent the widespread adoption of freeze-substitution.

Freeze-drying and freeze-substitution techniques need complementary staining procedures, for instance postfixation by OsO_4 or treatment with lead hydroxide, uranyl nitrate, phosphotungstic acid, or another contrast-enhancing reagent (for details, see 56, 90, 112). Cell components may also be visualized by a method requiring neither fixation nor embedding, namely the negative staining technique. Electron micrographs are obtained of the fixed or unfixed specimen present in a thin layer of concentrated electron stain such as potassium phosphotungstate (19, 32, 38, 46, 52, 53, 122). A negatively stained specimen, thin enough for high resolution, can be obtained in various ways, for instance, by spraying a suspension of the particles to be examined in a phosphotungstate solution on a carbon-covered grid or by spreading cell components on the surface of a drop of phosphotungstate solution ["surface spreading technique" (82)].

Substances used as embedding material for negative staining should be stable in water and nonreactive to cell substances at the pH used, apart from having a high electron density. Certain viruses retain their biological activity after phosphotungstate embedding. Apart from potassium phosphotungstate (*46*), various uranyl salts are employed— recently, for instance, uranyl oxalate (*73*). Objects amenable to negative staining should be of subcellular dimensions and have relatively stable structures such as microorganisms, viruses, and mitochondrial fragments. Apart from this limitation, negative staining is excellently suited to the study of freshly prepared biological materials.

From this brief survey of newer methods of specimen preparation, it should be evident that osmium tetroxide and permanganate fixation no longer dominate the field; on the other hand, these metal fixatives continue to perform an essential role and to find new applications. Elucidation of their chemical and physicochemical effects is useful not only in the development of new techniques but also in placing the interpretation of ultrastructure on a firm basis.

ACKNOWLEGMENTS

Support was received from the Dutch Organization for Pure Research (ZWO SON). The author is indebted to Dr. W. T. Daems for helpful comments.

REFERENCES

1. Adams, C. W. M., ed., "Neurohistochemistry," pp. 36 and 334. Elsevier, Amsterdam, 1965.
2. Adams, C. W. M. Osmium tetroxide and the Marchi method. *J. Histochem. Cytochem.* **8**, 262 (1960).
3. Afzelius, B. A. Chemical fixatives for electron miscroscopy. *Symp. Intern. Soc. Cell Biol.* **1**, 1–19 (1962).
4. Anderson, O. R., Roels, O. A., Dreher, K. D., and Schulman, J. H. The stability and structure of mixed lipid monolayers and bilayers. *J. Ultrastruct. Res.* **19**, 600 (1967).
5. Anderson, O. R., and Roels, O. A. Myelin-like configurations in Ochromonas malhamensis. *J. Ultrastruct. Res.* **20**, 127 (1967).
6. Bahr, G. F. Osmium tetroxide and ruthenium tetroxide and their reactions with biologically active substances. *Exptl. Cell Res.* **7**, 457 (1954).
7. Bahr, G. F. Continued studies about the fixation with osmium tetroxide. *Exptl. Cell Res.* **9**, 277 (1955).
8. Bahr, G. F., Bloom, G., and Friberg, U. Problems of osmium fixation. *Proc. Reg. Conf. (Eur.) Electron Microscopy, Stockholm, 1956* p. 106. Academic Press, New York, 1957.
9. Bahr, G. F., Bloom, G., and Friberg, U. Volume changes of tissues in physiological fluids. *Exptl. Cell Res.* **12**, 342 (1957).

10. Bahr, G. F., Bloom, G., and Johannisson, E. Further studies on fixation with osmium tetroxide. *Histochemie* **1**, 113 (1958).

11. Bahr, G. F., and Zeitler, E. H. Image interpretation and contrast in thin sections. *RCA Sci. Instr. News* **4**, 1 (1959).

12. Baker, J. R. "Cytological Technique," 4th ed. Methuen, London, 1960.

13. Barka, T., and Anderson, P. J. "Histochemistry," p. 136. Harper, New York, 1963.

14. Barrnett, R. J., and Roth, W. D. Effects of fixation on protein histochemistry. *J. Histochem. Cytochem.* **6**, 406 (1958).

15. Bayer, M., KMnO₄-Fixierung von Blutelementen. *Proc. 4th Intern. Conf. Electron Microscopy, Berlin, 1958* Vol. II, p. 29. Springer, Berlin, 1960.

16. Beer, M. Selective staining for electron microscopy. *Lab. Invest.* **14**, No. 6, Part 2, 282 (1965).

17. Birbeck, M. S. C., and Mercer, E. H. Applications of an epoxide embedding medium to electron microscopy. *J. Roy. Microscop. Soc.* [3] **76**, 59 (1956).

18. Birch-Anderson, A. The use of epoxy resins as embedding media. *Proc. 4th Intern. Conf. Electron Microscopy, Berlin, 1958* Vol. II, p. 44. Springer, Berlin, 1960.

19. Brenner, S., and Horne, R. W. A negative staining method. *Biochim. Biophys. Acta* **34**, 103 (1959).

20. Bullivant, S. Freeze-substitution and supporting techniques. *Lab. Invest.* **14**, No. 6, Part 2, 440 (1965).

21. Burkl, W., and Schiechl, H. A study of osmium tetroxide fixation. *J. Histochem. Cytochem.* **16**, 157 (1968).

22. Chapman, D., and Fluck, D. J. Physical studies of phospholipids. *J. Cell Biol.* **30**, 1 (1966).

23. Coleman, J. E., Ricciutti, C., and Swern, D. Improved preparation of keto-hydroxystearic acids. *J. Am. Chem. Soc.* **78**, 5342 (1956).

24. Criegee, R. Osmiumsäure-Ester als Zwischenprodukte bei Oxydationen. *Ann. Chem.* **522**, 75 (1936).

25. Criegee, R. Organische Osmiumverbindungen. *Angew. Chem.* **51**, 519 (1938).

26. Criegee, R., Marchand, B., and Wannowius, H. Zur Kenntnis der organischen Osmiumverbindungen. *Ann. Chem.* **550**, 99 (1942).

27. Dalton, A. J. A chrome-osmium fixative for electron microscopy. *Anat. Record* **121**, 281 (1955).

28. Doggenweiler, C. F., and Heuser, J. E. Ultrastructure of the prawn nerve sheath. *J. Cell Biol.* **34**, 407 (1967).

29. Elbers, P. F. Ion permeability of the egg of Limnaea Stagnalis. *Biochim. Biophys. Acta* **112**, 318 (1966).

30. Elworthy, E. H., and McIntosh, D. C. Micelle formation by lecithin in some aliphatic alcohols. *J. Pharm. Pharmacol.* **13**, 663 (1961).

31. Ericsson, J. L. E., and Biberfeld, P. Studies on aldehyde fixation. *Lab. Invest.* **17**, 281 (1968).

32. Fernández-Morán, H. New approaches in the study of biological ultra-structure" *Symp. Intern. Soc. Cell Biol.* **1**, 411–427 (1962).

33. Finean, J. B. Electron microscope and X-ray diffraction studies of a saturated synthetic phospholipid. *J. Biophys. Biochem. Cytol.* **6**, 123 (1959).
34. Finean, J. B. The molecular organization of cell membranes. *Progr. Biophys. Mol. Biol.* **16**, 143 (1966).
35. Finean, J. B., and Robertson, J. D. Lipids and the structure of myelin. *Brit. Med. Bull.* **14**, No. 3, 267 (1958).
36. Fleischer, S., Fleischer, B., and Stoeckenius, W. Fine structure of lipid-depleted mitochondria. *J. Cell Biol.* **32**, 1963 (1967).
37. Geren, B. B., and Schmitt, F. O. The structure of the nerve sheath in relation to lipid and lipid-protein layers, *J. Appl. Phys.* **24**, 1421 (1953).
38. Glauert, A. M. Factors influencing the appearance of biologic specimens in negatively stained preparations. *Lab. Invest.* **14**, No. 6, Part 2, 331 (1965).
39. Glauert, A. M. The fixation and embedding of biological specimens. *In* "Techniques for Electron Microscopy" (D. H. Kay, ed.), 2nd ed., pp. 166–212. Blackwell, Oxford, 1966.
40. Gottlieb, A., and Hecht, F. Colorimetrische Bestimmung von Mangan in Gläsern. *Mikrochemie* **35**, 337 (1950).
41. Griffith, W. P. Osmium and its compounds. *Quart. Rev. (London)* **19**, 254 (1965).
42. Hair, M. L. "Infrared Spectroscopy in Surface Chemistry," p. 166. Marcel Dekker, New York, 1967.
43. Hair, M. L., and Robinson, P. L. Reaction of ruthenium and osmium tetroxides with NH_3. *J. Chem. Soc.* p. 2775 (1960).
44. Hagström, L., and Bahr, G. F. Penetration rates of OsO_4. *Histochemie* **2**, 1 (1960).
45. Hake, F. Studies on the reactions of OsO_4 and $KMnO_4$ with amino acids, peptides, and proteins. *Lab. Invest.* **14**, No. 6, Part 2, 470 (1965).
46. Hall, C. E. Electron densitometry of stained virus particles. *J. Biophys. Biochem. Cytol.* **1**, 1, (1955).
47. Hanker, J. S., Kasler, F., Bloom, M. G., Copeland, J. S., and Seligman, A. M. Coordination polymers of osmium. The nature of osmium black. *Science* **156**, 1737 (1967).
48. Harwood, H. J. Reactions of the hydrocarbon chain of fatty acids. *Chem. Rev.* **62**, No. 2, 99 (1962).
49. Heun, F. A., Decker, G. L., Greenawalt, J. W., and Thompson, T. E. Properties of lipid bilayer membranes. *J. Mol. Biol.* **24**, 51 (1967).
50. Highton, P. J., Murr, B. L., Shafa, F., and Beer, M. Electron microscopic study of base sequence in nucleic acids. *Biochemistry* **7**, 825 (1968).
51. Hooghwinkel, G. J. M., and van Niekerk, H. P. G. A. Quantitative aspects of the tricomplex staining procedure. *Koninkl. Ned. Akad. Wetenschap., Proc.* **B63**, 358 (1960).
52. Horne, R. W. Negative staining methods. *In* "Techniques for Electron Microscopy" (D. H. Kay, ed.), pp. 328–355. Blackwell, Oxford, 1965.
53. Horne, R. W. The effects of negative stains. *Protoplasma* **63**, 212 (1967).
54. Ito, S. Post-mortem changes in the plasma membrane. *Proc. 5th Intern. Conf.*

Electron Microscopy, Philadelphia, 1962 Vol. 2, Art. L-5, Academic Press, New York, 1962.

55. Job, P. Recherches sur la formation de complexes. *Ann. Chim. (Paris)* [10] 9, 113 (1928).
56. Kay, D. H., ed. "Techniques for Electron Microscopy" 2nd ed. Blackwell, Oxford, 1965.
57. Khan, A. A., Riemersma, J. C., and Booij, H. L. The reactions with osmium tetroxide with lipids and other compounds. *J. Histochem. Cytochem.* 9, 560 (1961).
58. Klima, J. Fixierungs- und Einbettungsstudien für die Ultrahistologie. *Proc. 4th Intern. Conf. Electron Microscopy, Berlin, 1958* Vol. II, p. 58. Springer, Berlin, 1960.
59. Korn, E. D. II. Synthesis of bis (methyl 9,10-dihydroxy-stearate) osmate. *Biochim. Biophys. Acta* 116, 317 (1966).
60. Korn, E. D. III. Modification of oleic acid during fixation of amoebae by osmium tetroxide. *Biochim. Biophys. Acta* 116, 325 (1966).
61. Korn, E. D. The products of the reaction of osmium tetroxide with unsaturated lipids. *J. Cell Biol.* 34, 627 (1967).
62. Korn, E. D., and Weissman, R. A. I. Loss of lipids during preparations of amoebae for electron microscopy. *Biochim. Biophys. Acta* 116, 309 (1966).
63. Krames, B., and Page, E. Effects of electron microscopic fixatives. *Biochim. Biophys. Acta* 150, 24 (1968).
64. Latner, A. L., Molyneux, L., and Rose, J. D. A semiautomatic recording densitometer for use after paper-strip electrophoresis. *J. Lab. Clin. Med.* 43, 159 (1954).
65. Leduc, E. H., and Bernard, W. Water-soluble embedding media. *Symp. Intern. Soc. Cell Biol.* 1, 21–45 (1962).
66. Low, F. N. The electron microscopy of sectioned lung tissue. *Anat. Record* 120, 827 (1954).
67. Luft, J. H. Permanganate—a new fixative for electron microscopy. *J. Biochem. Cytol.* 2, 799 (1956).
68. Luft, J. H., and Wood, R. L. The extraction of tissue protein during and after fixation with OsO_4 in various buffer systems. *J. Cell Biol.* 19, 46A (1963).
69. Maan, C. J. The importance of the acetone and boric acid methods in the study of alicyclic 1,2-diols. *Rec. Trav. Chim.* 48, 332 (1929).
70. McLean, J. D. Fixation of plant tissue. *Proc. 4th Intern. Conf. Electron Microscopy, Berlin, 1958* Vol. II, p. 27. Springer, Berlin, 1960.
71. Maunsbach, A. B., Madden, S. C., and Latta, H. Variations in fine structure of renal tubular epithelium under different conditions of fixation. *J. Ultrastruct. Res.* 6, 511 (1962).
72. Meites, L. Polarographic characteristics of osmium. *J. Am. Chem. Soc.* 79, 4631 (1957).
73. Mellema, J. E., van Bruggen, E. F. J., and Gruber, M. Uranyl oxalate as a negative stain. *Biochim. Biophys. Acta* 140, 180 (1967).

74. Mellor, J. W. "A Comprehensive Treatise on Inorganic and Theoretical Chemistry," p. 265. Longmans, Green, New York, 1932.
75. Millonig, G. Model experiments on fixation and dehydration. *Proc. 6th Intern. Conf. Electron Microscopy, Kyoto, Japan, 1966* Vol. 2, p. 21. Maruzen Co., Ltd., Tokyo, 1967.
76. Millonig, G., and Marinozzi, V. Fixation and Embedding in electron microscopy. *In* "Advances in Optical and Electron Microscopy" (V. E. Cosslett and R. Barer, eds.), Vol. 2, p. 251. Academic Press, New York, 1968.
77. Mühlethaler, K. Die Dehydratisierung. *Proc. 4th Intern. Conf. Electron Microscopy, Berlin, 1958* Vol. II, p. 32. Springer, Berlin, 1960.
78. Nageotte, J. "Morphologie des Gels Lipoides." Hermann, Paris, 1936.
79. North, W. J. Method for revealing the membrane systems in micro-organisms. *Nature* **190,** 1215 (1961).
80. Palade, G. E. A study of fixation for electron microscopy. *J. Exptl. Med.* **95,** 285 (1952).
81. Palade, G. E. The fixation of tissues for electron microscopy. *Proc. 3rd Intern. Conf. Electron Microscopy, London, 1954* p. 129. Roy. Microscop. Soc., London, 1956.
82. Parsons, D. F. Effects of the preparation procedures. *Lab. Invest.* **14,** No. 6, Part 2, 434 (1965).
83. Pearse, H. E. "Histochemistry Theoretical and Applied," 2nd ed., p. 59. Churchill, London, 1960.
84. Pease, D. C. "Histological Techniques for Electron Microscopy," 2nd ed. Academic Press, New York, 1964.
85. Pease, D. C. Eutectic ethylene glycol and pure propylene glycol as substituting media for the dehydration of frozen tissue. *J. Ultrastruct. Res.* **21,** 75 (1967).
86. Pease, D. C. The preservation of tissue fine structure during rapid freezing. *J. Ultrastruct. Res.* **21,** 98 (1967).
87. Porter, K. R., and Kallman, F. The properties and effects of osmium tetroxide. *Exptl. Cell Res.* **4,** 127 (1953).
88. Pourbaix, M. "Atlas of Electrochemical Equilibria in Aqueous Solutions," pp. 370–371. Pergamon Press, Oxford, 1966.
89. Rebhun, L. I. Freeze-substitution. *Federation Proc.* **24,** Suppl. 15, S217 (1965).
90. Reimer, L. "Elektronemikroskopische Untersuchungs- und Präparationsmethoden," 2nd ed. Springer, Berlin, 1967.
91. Revel, J. P., Ito, B., and Fawcett, D. W. Electron micrographs of myelin figures. *J. Biophys. Biochem. Cytol.* **4,** 495 (1958).
92. Revel, J. P., and Ito, S. The surface components of cells. *In* "The Specificity of Cell Surfaces" (B. D. Davis and L. Warren, eds.), p. 211. Prentice-Hall, Englewood Cliffs, New Jersey, 1967.
93. Rhodin, J. Correlation of ultrastructural organization and function in normal and experimentally changed proximal convoluted tubule cells of the mouse kidney. Thesis, Stockholm, Karolinska Institute, Aktiebolaget Godvil (1954).

94. Riemersma, J. C. Osmium tetroxide fixation of lipids: Nature of the reaction products. *J. Histochem. Cytochem.* **11**, 436 (1963).
95. Riemersma, J. C. Osmium tetroxide fixation of lipids: A possible reaction mechanism. *Biochem. Biophys. Acta* **152**, 718 (1968).
96. Riemersma, J. C., and Booij, H. L. The reactions of osmium tetroxide with lecithin: Application of staining procedures. *J. Histochem. Cytochem.* **10**, 89 (1962).
97. Robertis, E. de, and Lasansky, A. Ultrastructure and chemical organization of photoreceptors. *In* "The Structure of the Eye" (G. K. Smelser, ed.), pp. 29–49. Academic Press, New York, 1961.
98. Robertson, J. D. The molecular structure and contact relationships of cell membranes. *Progr. Biophys. Biophys. Chem.* **10**, 343 (1960).
99. Robertson, J. D. The molecular biology of cell membranes. *Mol. Biol. Symp., New York 1958* pp. 87–151. Academic Press, New York, 1960.
100. Rosenbluth, J. Contrast between osmium-fixed and permanganate-fixed toad spinal ganglia. *J. Cell Biol.* **16**, 143 (1963).
101. Ruff, O., and Bornemann, F. Ueber das Osmium. *Z. Anorg. Chem.* **65**, 429 (1910).
102. Ryter, A., and Kellenberger, E. Inclusion au polyester. *Proc. 4th Intern. Conf. Electron Microscopy, Berlin, 1958* Vol. II, p. 52. Springer, Berlin, 1960.
103. Ryter, A., and Kellenberger, E. Etude au microscope électronique de plasmas contenant de l'acide desoxyribonucleique. *Z. Naturforsch.* **13b**, 597 (1958).
104. Sabatini, D. D., Bensch, K. G., and Barrnett, R. J. New fixatives for cytological and cytochemical studies. *Proc. 5th Intern. Conf. Electron Microscopy, Philadelphia, 1962* Vol. 2, Art. L-3. Academic Press, New York, 1962.
105. Sabatini, D. D., Bensch, K. G., and Barrnett, R. J. Cytochemistry and electron microscopy. *J. Cell Biol.* **17**, 19 (1963).
106. Sauerbrunn, R. D., and Sandell, L. B. The ionization constants of osmic (VIII) acid. *J. Am. Chem. Soc.* **75**, 4170 (1953).
107. Schecter, A. Variations in unit membrane structure. *Proc. 6th Intern. Conf. Electron Microscopy, Kyoto, Japan, 1966* Vol. 2, p. 397. Maruzen Co., Ltd., Tokyo, 1967.
108. Schidlovsky, G. Contrast in multilayer systems after various fixations. *Lab. Invest.* **14**, No. 6, Part 2, 475 (1965).
109. Seligman, A. M., Wasserkrug, H. L., and Hanker, J. S. A new staining method (OTO) for enhancing contrast of lipid-containing membranes. *J. Cell Biol.* **30**, 424 (1966).
110. Sjöstrand, F. S. Critical evaluation of ultrastructural patterns with respect to fixation. *Symp. Intern. Soc. Cell Biol.* **1**, 47–68 (1962).
111. Sjöstrand, F. S. Molecular structure and function of cell membranes. *Protides Biol. Fluids Proc. Colloq.* **15**, 15 (1967).
112. Sjöstrand, F. S. "Electron Microscopy of Cells and Tissues," Vol. 1. Academic Press, New York, 1967.
113. Sjöstrand, F. S., and Elfvin, L. G. The granular structure of mitochondrial membranes. *J. Ultrastruct. Res.* **10**, 263 (1964).

114. Small, D. M. A classification of biologic lipids. *J. Am. Oil Chemistry Soc.* **45**, 108 (1968).

115. Stoeckenius, W. An electron microscope study of myelin figures. *J. Biophys. Biochem. Cytol.* **5**, 491 (1959).

116. Stoeckenius, W. Some electron microscopical observation on liquid-crystalline phases in lipid-water systems. *J. Cell Biol.* **12**, 221 (1962).

117. Stoeckenius, W., and Mahr, S. Studies on the reaction of OsO₄ with lipids and related compounds. *Lab. Invest.* **14**, No. 6, Part 2, 458 (1965).

118. Tooze, J. Measurements of some cellular changes during the fixation of amphibian erythrocytes with OsO₄ solution. *J. Cell Biol.* **22**, 551 (1964).

119. Tormey, J. M. Artifactual localization of ferritin. *J. Cell Biol.* **25**, No. 2, 1 (1965).

120. Trump, B. F., Young, D. E., Arnold, E. A., and Stowell, R. E. Effects of freezing and thawing on cytoplasmic structures. *Federation Proc.* **24**, Suppl. 15, S144 (1965).

121. Trump, B. F., and Ericsson, J. L. E. The effect of the fixative solution on the ultrastructure of cells and tissues. *Lab. Invest.* **14**, No. 6, Part 2, 507 (1965).

122. Valentine, R. C., and Horne, R. W. An assessment of negative staining techniques. *Symp. Intern. Soc. Cell Biol.* **1**, 263–278 (1962).

123. Wallach, D. F. H., and Gordon, A. Protein conformations in cellular membranes. *Protides Biol. Fluids, Proc. Collog.* **15**, 47 (1967).

124. Waters, W. A. Mechanisms of oxidation by compounds of chromium and manganese. *Quart. Rev. (London)* **12**, 277 (1958).

125. Weissenfels, N. Präparatveränderungen während der Fixierung. *Proc. 4th Intern. Conf. Electron Microscopy, Berlin, 1958* Vol. II, p. 60. Springer, Berlin, 1960.

126. Wetzel, B. K., Sodium permanganate fixation for electron microscopy. *J. Biophys. Biochem. Cytol.* **9**, 711 (1961).

127. Wolman, M. The reaction of osmium tetroxide with tissue components. *Exptl. Cell Res.* **12**, 231 (1957).

128. Wood, R. L., and Luft, J. H. The influence of buffer systems on fixation with osmium tetroxide. *J. Utrastruct. Res.* **12**, 22 (1965).

129. Zeiger, K. Probleme der Fixation in Licht- und Elektronemikroskopie. *Proc. 4th Intern. .Conf. Electron Microscopy, Berlin,* 1958 Vol. II, 17. Springer, Berlin, 1960.

130. Zetterqvist, H. The ultrastructural organization of the columnar absorbing cells of the mouse jejunum. Thesis Karolinska Institutet, Stockholm, Aktiebolaget Godvil (1956).

Chapter III

PRESENT STATUS OF FREEZING TECHNIQUES

STANLEY BULLIVANT

I. Introduction

Although there have been recent advances in technique which may eventually make it possible to examine cells in the electron microscope in the hydrated living state (*25, 32, 33*), these are only applicable in specialized situations and the majority of work is still carried out on thin, dehydrated specimens. This is a severe limitation, for it means that we do not have the image of the living cell at the ultra-structural level to use as a comparison standard, whereas light micros-copists do with phase microscopy of living material. The only logical way seems to be to examine a particular system by a whole variety of methods and if its structure correlates well from one method to the other and does not conflict with information obtained by light microscopy, diffraction studies, and biochemistry, then we can be reasonably sure that the structure we see is that possessed *in vivo*. A good example of this approach is provided by the mitochondrion. No one now seriously doubts that its interior is occupied by a series of infolded lipoprotein cristae. There are, however, many cellular organelles whose structure in the living state is still in doubt. For example, although the majority of evidence is in favor of ribosomes as small particles of about 150 Å diameter, they are not seen as such after permangate fixation (*68*), some methods of freeze-drying (*54, 114*) and after negative staining by the spread cell method (*95*). As new techniques for biological electron microscopy are brought into use, they are usually described as producing "less artifact" than previous methods. This is rather a subjective evaluation. The structure of proteins, nucleic acids, and lipids depends on their interaction with the surrounding layers of water, and the configuration of such large molecules is changed when the water is replaced by a nonpolar embedding medium prior to thin sectioning or by an amorphous glass of phosphotungstic acid as is the case in negative staining. An ideal solution would be the examination of cells in the hydrated state. The only way that this could be done for the majority of cells would be to freeze them rapidly and then cut sections of the cells embedded in ice in a manner analogous to the use of the cryostat microtome. These would be examined at low temperature in the electron microscope to insure that there was neither melting nor recrystallization. This will be dealt with in more detail later, and it will be seen that, although this ideal method is far from being fully developed, there have been some exciting advances recently. Even if cells are frozen without ice crystals—sectioned in ice and examined frozen—the results are still one step removed from *in vivo* conditions by the physical and chemical effects of the conversion of the water of the cells into ice. In the following sections the various methods of examining frozen material in the electron microscope will be discussed. Particular attention will be paid to the techniques of freeze-etching and freeze-fracturing which have recently come to the forefront, both in providing new information and in provoking controversy over the structure of biological material.

II. Physical Basis of Freezing Techniques

The mechanism of freezing in biological systems has been investigated at all levels from that of simple salt solutions, through more complex models of cytoplasm, the cell itself, the organ, and finally the whole organism. These investigations have been largely conducted from the point of view of survival of viability. Much of this work is beyond the scope of this present chapter, but a description of the effects of freezing on cells which are pertinent to their preservation for electron microscopy will follow. Several symposia and discussions on freezing in biology have been published over the years (*55, 56, 79, 93, 94, 116, 129*). The extensive work of Luyet and his group has been largely published in the journal *Biodynamica*. The book "Cryobiology" (*80*) (Meryman, ed.) is particularly informative. To avoid confusion, it should be noted that this book (*80*), the proceedings of a conference (*129*), and a journal all have the same title of "Cryobiology."

What happens when aqueous solutions freeze depends greatly on the speed of cooling. If water is cooled slowly and there are suitable nucleation centers available, it turns into crystalline ice at 0°C, and the whole volume remains at this temperature until all the water has been converted to ice. The faster the cooling the smaller the ice crystals formed. If very pure water is cooled, particularly in a finely divided state, the water may supercool as low as −40°C. Induced or spontaneous nucleation with ice causes very rapid freezing of such supercooled water. A particularly important fact to keep in mind is that ice is not a static structure. At all temperatures between 0°C and −130°C it is subject to migratory recrystallization with large ice crystals growing at the expense of small ones because of surface energy considerations (*78, 82*). Investigators, using freezing as a technique for cytological studies, have very often claimed that they froze the material so fast that it became vitreous (amorphous or glassy). From this point of view vitrification has been defined operationally; the ice crystals formed are too small to see. Material frozen fast enough to appear vitreous in the light microscope may turn out to have submicronic ice crystals when seen in the electron microscope or when studied by x-ray diffraction. Pryde and Jones (98) claimed to produce vitreous water by deposition from the vapor at temperatures below $-130°C$. On warming, they observed an evolution of heat at $-129°C$, which they interpreted as being due to the crystallization of water. Lonsdale (65) believes that x-ray diffraction patterns of ice deposited from the vapor at temperatures below $-130°C$ indicate that the ice is vitreous rather than crystalline. If one interprets vitrification as the production of ice crystals too small to interfere with resolution in electron microscope specimens, then it will be seen that it is attainable. On the other hand, if one adopts a more rigorous definition that requires cooling to well below $-100°C$ in a time which is short compared with the time in which water molecules rearrange themselves, that is of the order 10^{-12} seconds, then it appears impossible to attain experimentally (52). From the standpoint of cytology at present, this argument is academic; most important, the ice crystals must not interfere with the image.

If salt is added to the water there is the additional complication of the formation of eutectics. The eutectic temperature is defined as the lowest temperature at which a solution of the salt can remain in equilibrium with ice. The eutectic concentration is the concentration of the still-dissolved salt at this point. In the case of NaCl, the eutectic temperature is $-21°C$ and the concentration is 30% (5.2 M). When a dilute solution of NaCl is frozen, water initially crystallizes out to give pure ice without inclusion of solute. This increases the concentration of the remaining salt solution; water continues to crystallize out as pure ice until the eutectic temperature and concentration are reached. The temperature stays at this point until the remaining water and the salt have formed a crystalline hydrate, $NaCl \cdot H_2O$; further removal of heat causes the temperature of the system to be lowered. Most eutectics other than that formed by NaCl crystallize out as the salt and ice. The importance of eutectic formation is two-fold: It leads to damaging concentrations of electrolytes and because the different components of buffer systems have different eutectic temperatures, there may also be damaging pH changes (usually toward the acid side) on freezing. The above description is not completely true for it is often found in practice that there is considerable supercooling of eutectics. This extends their damaging effects to lower temperatures. Protoplasm, which contains proteins, lipids, and nucleic acids, in addition to a number of electrolytes, has an ill-defined eutectic temperature. This means that ions will be mobile and there will be osmotic and other effects of high salt concentrations down to temperatures as low as —45°C (*108*). Protoplasm has the advantage over pure water that ice crystal growth is slowed by the presence of macromolecules and solutes and smaller crystals result at similar cooling rates. Except for a few which have a favorable size and shape, ions in general are not incorporated into ice to give mixed crystals. There is good evidence that if protoplasm is frozen rapidly (several thousand degrees centigrade per second) ice crystals, below the size at which they interfere with electron microscopy, will be formed. Salts and macromolecules will be forced out into the interstices between the crystals. The temperature above which migratory recrystallization takes place in frozen protoplasm is probably considerably above the —130°C previously quoted for pure water (*82*) on account of the hindrance caused by the frozen-out solutes.

We now come to the problem of cell freezing. As already mentioned, most investigations have been directed at obtaining cell survival. Suspensions of cells which have not been protected with an agent such as glycerol show good recovery after freezing and thawing when the freezing has taken place either slowly or very rapidly (*74*). There is a region of low recovery between these two peaks and the position of the peaks depends on the cell type, being about 1°C per minute and above 1000°C per minute for yeast. Mazur's interpretation (*74*) is as follows: Cells frozen at 1°C per minute are shrunk osmotically as a result of the formation of large extracellular ice crystals. There is no intracellular ice as only bound water is left in the cells. This conclusion is supported by electron micrographs (*99*). At freezing rates higher than this, relatively large intracellular ice crystals are formed and these are damaging (see Fig. 1). At high freezing rates the intracellular ice crystals formed are so small as not to be damaging as long as thawing is carried out rapidly so that the crystals do not grow appreciably during warming. Cell death can be caused either by large intracellular ice crystals or by exposure to high salt concentrations resulting from eutectic formation. The mechanism in either case is not fully understood, but it appears that freezing damage is not simply mechanical.

For morphological examination in the electron microscope rapid

freezing methods have been used. These methods will be described
in the review of the various methods of processing frozen specimens
in the next section. Cells frozen slowly without intracellular ice crystal
formation are so shrunken that the morphology is difficult to distinguish (99).

A method which has been equally valuable for survival and morphological studies is the use of glycerol as a protective agent (66).
Other agents, such as dimethyl sulfoxide, have a similar effect (67).
Most cell types do not show any appreciable survival without the
use of a protective agent. Glycerol passes freely through cell membranes and for each molecule of glycerol three of water are bound.
It acts by making some water unavailable for freezing and by diluting
solutes and avoiding damaging concentrations. Its action in giving
good morphological preservation seems to be that it effectively de-
hydrates the cell and makes less water available for freezing. Sufficiently small ice crystals can be obtained with much slower cooling.
The use of glycerol lowers the eutectic point of whole tissue from
−45° to −60°C (108). When glycerol solutions are frozen, the ice
crystallizes out until the remaining glycerol reaches a concentration
of 60–70% and then this solidifies into a glass (81).

III. Processing for Examination

A. WHOLE-MOUNT PREPARATIONS

Early work on the freezing of whole-mount preparations was restricted to small particles such as bacteria and viruses. A very good
review of the methods was published by Williams (140). His method
involved spraying very small droplets containing the particles under
investigation onto a thin collodion film in intimate contact with a
copper block held at liquid nitrogen temperature. Using this method
it was calculated that with a drop that flattens on impact to a thickness of 4 μ, cooling rates as high as 10^6 °C per second might be obtained. Freeze-drying was carried out and heavy metal shadowing
done in situ on the collodion film, which was then stripped off and
examined in the electron microscope. Williams recommends the use
of suspension media containing volatile electrolytes as these will not

interfere with the final image. The method gives very good pictures of a variety of biological objects and demonstrates that they have not been subject to the surface tension forces that produce flattening of air dried objects. Although the spray freeze-dry method is not used much at present, it does appear that it would be valuable for very rapid freezing of suspensions of such objects as mitochondria or chromosomes, followed by freeze-drying, freeze-substitution, or freeze-etching.

A recent example of whole-mount techniques is the replication of the surfaces of freeze-dried tissue culture cells to reveal the microvilli extending from them (*40*).

Freeze-drying is also used to prepare specimens for scanning electron microscopy, but this is outside the scope of this review.

B. Thin Sections

1. *Freeze-Drying*

Ice has quite a high vapor pressure even as low as $-70°C$ (2×10^{-3} mm Hg), and if it is kept at this low temperature in a high vacuum and in close proximity to a cold trap at a lower temperature, the ice will sublime away. This is the basis of freeze-drying. The object of freeze-drying and freeze-substitution techniques has in general been to preserve labile substances within cells and to maintain intracellular relationships which conventional methods might destroy.

Freeze-drying as practiced for electron microscopy is an extension of the same method used for light microscopy, the differences being due to the higher resolving power and specimen requirements of the electron microscope. The main requirement is that much smaller ice crystals be produced in the specimen. Light microscopists are not troubled by crystals below about $\frac{1}{2}\ \mu$ in size, the resolving limit of their instrument, although crystals $100\ \text{Å}$ across could be very disturbing in the electron microscope. The search has always been for very high cooling rates to produce very small crystals or amorphous ice. Measurements on a variety of coolants have been done using thermocouples embedded in small specimens (*15, 102, 126*). Propane gives the most rapid cooling although the liquid Freons are comparable. Cooling rates in excess of $2000°C$ per second are attainable with specimens of the size used in electron microscopy (of the order of 1 mm on edge). When a specimen at room temperature is immersed in liquid nitrogen an insulating layer of vapor is formed around it and the specimen cools relatively slowly until the temperature differential between it and the nitrogen is insufficient to support the layer. There is only rapid cooling for the last few degrees. Helium I and helium II behave in a similar manner but the insulating layer is not so persistent in propane (15). It is to be expected that propane would be better than liquid nitrogen for it has a much higher boiling point, $-40°C$ as opposed to $-196°C$. The ideal coolant would be one with a boiling point above room temperature and a freezing point as low as possible combined with low viscosity near that point and having high thermal conductivity and heat capacity. Propane, isopentane, Freon 12, and Freon 22 are nearer to this ideal than liquid gases such as nitrogen. When using liquid propane, it is advisable to avoid letting its temperature fall below $-175°C$ as there is the possibility that oxygen out of the air may condense in it and produce an explosive mixture (125). It is safer to use the Freons. Other methods of obtaining rapid cooling will be discussed in connection with the work of the investigators who used them.

The theory and method of freeze-drying is discussed in the reviews cited at the beginning of this chapter. For details Malmstrom (73), Glick and Malmstrom (48), Gersh and Stephenson (44) and Stephenson (127) should be consulted. The recent work is covered in Meryman's review (81). The tissue being dried should be held at $-70°C$, which is below the eutectic point of any component. There is little point in going below this because it only results in extremely long drying times. There is general agreement that a drying chamber pressure of 10^{-5} mm Hg is adequate, and Gersh and Stephenson (44) state that 10^{-4} mm Hg is always adequate regardless of drying temperature. The pumping system keeps the air pressure in the chamber low enough so as not to interfere with drying and a cold trap is inserted to remove the water and protect the pumps. At liquid nitrogen temperature, the cold trap is always capable of removing the water. As the specimen dries, a shell of dry material is formed around its periphery; this reduces the flow of water molecules outward. The water vapor pressure at the ice interface within this shell is the vapor pressure of water in equilibrium with ice at the temperature of the specimen. Stephenson (*127*) has pointed out that this shell of dried material constitutes a major barrier to the flow of vapor; thus, there is no point in making other parts of the system able to remove vapor at a much greater rate than it can flow through this shell. This point is adequately brought out in a practical manner by the large number of designs of freeze-drying apparatus which have been found almost equally successful. A standard design for freeze-drying equipment is that of Glick and Malmstrom (*48*). Using a specimen temperature of $-80°C$, a chamber pressure of 10^{-5} mm Hg, and a liquid nitrogen cold finger, they were able to dry a 2-mm cube of tissue in 6 hours. Most designs require that the specimen temperature be kept constant; this was achieved in their case by keeping the specimen-containing part of the chamber in a dry ice–acetone bath. Other designs keep the whole chamber at liquid nitrogen temperature and heat the specimen electrically via a feedback controlled system actuated by a thermocouple in the specimen. An example of this type is the equipment of Stirling and Kinter (*128*). Once the initial drying at low temperature is over, a further period of secondary drying at a higher temperature is usually carried out. The foregoing remarks on freeze-drying apply equally well to light or electron microscopy but lower specimen temperatures during drying are the rule for the latter. Particular points of technique pertinent to electron microscopy will be brought out in the following review.

The work on freeze-drying for electron microscopy has mainly stemmed from two sources under the inspiration of Gersh and Sjöstrand.

Gersh's work (*41*) made freeze-drying of great practical value for light microscopy, and more recently he and a group of his colleagues have explored it both theoretically and practically from the standpoint of electron microscopy (*42, 44–47, 126*). After drying, they fixed and stained the proteins with alcoholic heavy metal-containing solutions. Histochemical studies with the electron microscope have been done using Gersh's freezing technique by Bondareff (*5*) for glycogen and Finck (*39*) for basophilic substances. This group has used drying temperatures of $-35°$ to $-40°C$ which seems rather high when it is considered that it is higher than the eutectic point of tissue (*108*). That the published micrographs all show structure seemingly disrupted by ice crystals is the main criticism of this work. Gersh (*43*) believes that this is the true structure, but others, notably Rebhun (*102*), do not agree. Mundkur (*92*) has done an extensive investigation of the histochemistry of freeze-dried yeast cells using modifications of Gersh's methods. He used the vapor of the protein cross-linking agent difluorodinitrobenzene (*47*) to stabilize the cells after drying. He also used a somewhat lower drying temperature. His preparations are apparently ice crystal free.

Sjöstrand (*111, 112*) did significant early work on the electron microscopy of frozen dried material. Müller's study (*91*) of plant material introduced better ways of embedding frozen tissue. A number of studies of the pancreas by Sjöstrand and those inspired by his work followed on these improvements (*53, 54, 114*). This group has usually used anhydrous osmium tetroxide vapor to stabilize the dried tissue after it has been warmed to room temperature from the drying temperature of $-70°C$, and before it is embedded by vacuum impregnation. After treatment in this way, membranes appear in positive contrast, but appear as negative images if the osmium step is omitted. Their most important conclusion was that the ribosomes seen in chemically fixed material were artifacts in so far as they were aggregated into particulate form (see section on freeze-substitution for counterarguments). Grunbaum and Wellings (*49*) also claim that the particulate ribosomes are artifacts. Elfvin (*27*) has done a high resolution study of membranes and demonstrated that after osmium tetroxide vapor stabilization and on-section staining they have the familiar unit membrane structure, but with somewhat larger dimensions (95 Å overall). Sjöstrand and Elfvin (*115*) have studied mitochondrial and other cytoplasmic membranes after freeze-drying and conclude that a granular appearance seen in the central less dense lamina of the unit membrane reflects a "globular substructure consisting of the membrane lipids in micellar arrangement, stabilized by proteins." This is in support of Sjöstrand's membrane model (*113*), based on earlier work with chemical fixation. Seno and Yoshizawa (*109*), using a similar technique to the Sjöstrand school, but without the vapor stabilization, saw ribosomes as particles and even claimed to remove them with RNase.

Most exponents of freeze-etching (with the exception of Gersh's group) have shown ice crystal free preparations but have admitted that they did have ice crystals in other parts of their specimens.

The ultimate aim of freeze-drying methods is to leave small molecules and ions in place so that they can be detected by cytochemical or autoradiographic methods. Several of the applications cited have employed cytochemical methods on the sections. The difficulty with autoradiography is to avoid penetration of water and leaching during thin sectioning and during the photographic process. An ingenious method of overcoming this difficulty has recently been described by Stirling and Kinter (*128*). These workers froze small pieces of hamster intestine in propane at $-184°C$ and then used a freeze-drying schedule extending from $-70°C$ to room temperature over a period of four days. The tissue was fixed under vacuum with the vapor of anhydrous osmium tetroxide and finally embedded in Araldite containing 1% silicone fluid. The sections were cut on water, dried, coated with a liquid photographic emulsion, exposed, and developed. The silicone prevented the wetting of small channels within the tissue; tritiated galactose did not leach out and could be located autoradiographically fairly sharply within a $2\text{-}\mu$ band. Most of the work concerned light microscopy but some electron microscope autoradiographs were presented and demonstrated the feasibility of the technique. These showed many small ice crystals and this is to be expected considering that no protective agent was used during freezing.

2. *Freeze-Substitution*

Freeze-substitution, introduced by Simpson (*110*) in 1941, surprisingly has not enjoyed as great a popularity as freeze-drying. It is much simpler to do and uses less elaborate apparatus, although it does have the disadvantage that it takes rather a long time for substitution by the solvent to occur at temperatures as low as $-80°C$. Work such as that of Feder and Sidman (*28*) shows the value of the technique for light microscopy. The remarks on methods of obtaining high cooling rates in the section on freeze-drying apply equally well to freeze-substitution. If a dry solvent such as alcohol or acetone is cooled to the temperature used for freeze-substitution it still has an appreciable ability to dissolve water. In the case of acetone, Van Harreveld *et al.* (*131*) found, by nuclear magnetic resonance experiments, that water is taken up to the extent of 2% at $-85°C$. When a piece of frozen tissue is immersed in acetone at this temperature the ice is dissolved in the acetone molecule by molecule, whereby it effectively changes from ice to dissolved water, having lost the constraints of the ice structure. There is no reason why this method of water removal should be any more disruptive than freeze-drying. At this low temperature the acetone will have little chemical reaction with the tissue. At the interface between the ice and the acetone, the maximum local dissolved water concentration will be 2%; at −85°C NaCl has a solubility of only 0.0018% (*131*). This means that there will be little diffusion of salts and presumably even less of larger molecules which can be expected to be relatively less soluble.

Fernández-Morán has been aware of the possibilities of cryofixation in electron microscopy for a long time. His first work in the field was in 1952 (*29, 38*). He initiated the application of freeze-substitution to electron microscopy with some experiments using glycerinated tissue (*30*). Later (*31, 32*) he introduced liquid helium II as a coolant in the belief that its super heat conductivity and super fluidity would give high cooling rates. Although it now seems that this is not the case (*15*), at least for specimens as large as are commonly embedded for electron microscopy, his dramatic approach to the problem engendered great interest in cryofixation and gave the impetus which started other people in the field. Fernández-Morán has published a number of papers which describe his work in low temperature electron microscopy (*29–38*). He used either fresh tissue or protected it by glycerination, froze it in liquid helium II or liquid nitrogen–Freon mixtures, substituted in a variety of solvents with the addition of heavy metal salts, and embedded in methacrylate by photopolymerization. He demonstrated very good preservation of lamellar systems such as retinal rod outer segments and myelin sheath and found that glycerination was essential for ice-crystal-free preparations of larger specimens. Apart from the work on freeze-substitution he has been active in all aspects of low temperature electron microscopy. In particular, he has been interested in using low temperature stages for examination of specimens in the frozen state. A complete review of all his work is beyond the scope of this chapter and his publications should be consulted.

Rebhun (*101*) started to use freeze-substitution on marine eggs because he was discouraged with the results of chemical fixation. Although he obtained good preparations only in about 5% of the cases, the preservation in this 5% was excellent, with unbroken membrane systems and unextracted background. He found ribosomes along the endoplasmic reticulum and by analogy with the fact that other facets of cell fine structure in his material also depended for their visualization on the stain used, he argued that Hanzon *et al.* (*54*) were incorrect in dismissing the particulate ribosomes as artifacts. Disturbed by the low percentage of success, he did a very careful study of freezing and freeze-substitution and put forward the hypothesis that some form of protection, either by glycerination or accidental drying was essential to obtain ice crystal free preparations (*103*). A fuller description of his experiments on the relation between the fine structure of cells after freeze-substitution and the water concentration at the time of freezing is given in a later paper (*102*). Although he found that propane and propylene were the best of a wide range of coolants tested, he used either Freon 22 or Genetron 23 as these produced only slightly lower cooling rates and were much safer to use. He describes five major types of morphology depending on the water content of the frozen cells, ranging from large ice crystals in cells which have not lost water, through morphologies rather similar to those seen after chemical fixation, to shrunken cells resulting from extreme dehydration. He prefers the picture seen after freezing with a water content of 40–50% below normal. This is similar to chemically fixed material and has well stained nuclei with distinct blocks of chromatin and clearly visible ribosomes; after suitable staining, the unit membranes with their triple layered structure are seen. A particularly interesting point is that the ribosomes only appear at this intermediate level of dehydration and are not seen when the normal amount of water is present or after extreme dehydration. It is possible that the ribosomes are obscured by ice crystals in cells frozen with their normal water content.

The earlier work of the author (*12, 13*) employed either helium II or propane as a coolant followed by freeze-substitution in studies on mouse pancreas. Membranes appeared in negative contrast, particularly after heavy metal staining. The ribosomes were not easily visible in unstained material but were revealed by heavy metal staining and were found to be somewhat variable in size and in more intimate contact with a dense layer on the membrane surface than is usual in chemically fixed material.

The percentage of success was rather low, being similar in retrospect to the experience of Rebhun and Gagne (*103*). As it is now known that helium II is a relatively poor coolant (*15*), the reason may have been the same, namely, that the infrequent good blocks were due to accidental dehydration before freezing. It was found that similar preservation to that obtained occasionally with helium II could be produced routinely by glycerination and freezing in propane at −175°C (*13, 15*). [Full details of the method are given in the later publication (*15*).] Material prepared in this way has been used to study differential staining of nuclear structures with heavy metal salts (*15, 132*) and the problem of negative images of membranes in frozen material (*14, 15*). Figures 1 and 2 demonstrate the effectiveness of glycerol in protecting cells against ice crystal damage. Figure 1 shows a tissue culture cell after direct freezing in propane. It is

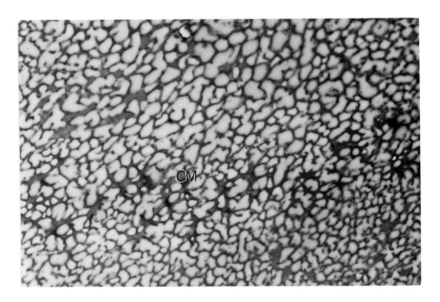

FIG. 1. Metaphase FL tissue culture cell prepared by freezing in propane, substituting in ethanol containing 1% OsO_4 at −75°C, embedding in methacrylate, and on-section staining with osmium. Relatively large ice crystals, seen as empty "ghosts," disrupt the structure. The remainder of the cytoplasm appears as a network representing the eutectic. The chromosomes (CM) of the metaphase plate are quite distorted. 15,000×.

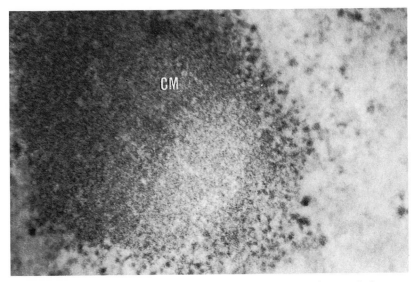

Fig. 2. Chromosome (CM) in metaphase FL cell. Preparation made by pretreating cells with a graded glycerol series up to 60%, freezing in propane, substituting at $-75°C$ in ethanol, embedding in Epon, and on-section staining with lead hydroxide. The use of glycerol has prevented ice crystal formation and the fibrillar structure of the chromosome is preserved. 84,000×.

filled with relatively large ice crystals. A portion of a similar cell which was pretreated with glycerol is shown in Fig. 2. There is no visible ice crystal damage. It is of interest that it was never possible to obtain good preservation of unprotected tissue culture cells, although occasionally it was with the pancreas. This may be related to the higher water content of the tissue culture cells. Mouse pancreas which is glycerinated shows large well preserved areas (*15*). Membranes appear in negative contrast in frozen material if the tissue is put into the embedding medium before it is stained. Such "negative membranes" as seen in mitochondria and the Golgi region are shown in Figs. 3 and 4. The white line is 30–40 Å across; it is presumed that it is due to the central, less dense lamina of the unit membrane, with additional thickness due to a region on either side which does not stain. This region is probably that occupied by the hydrophilic ends of the lipid molecules. The protein on the surface of the mem-

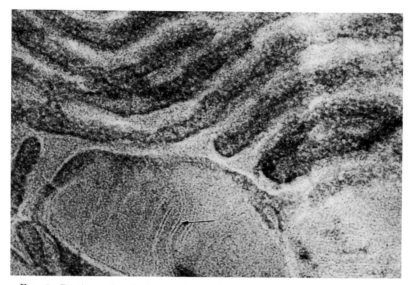

Fig. 3. Portion of cytoplasm of exocrine cell of mouse pancreas prepared by treating with a graded glycerol series up to 60%, freezing in propane, substituting in ethanol at −75°C, embedding in methacrylate, and on-section staining with potassium permanganate. The mitochondrial cristae (arrow) appear as two unit membranes back-to-back with the inner layers fused to give a dense line. 120,000×.

brane is not visible because it has the same contrast as the surroundings. After suitable staining a unit membrane structure can be seen (*14, 101, 102*). This is illustrated in Fig. 3 where the inner layer of the two back-to-back membranes of the mitochondrial cristae is much more dense than the matrix after potassium permanganate staining. The good preservation illustrated is not dependent on the presence of osmium tetroxide or any other fixative—a point of some importance. The addition of osmium tetroxide to the substituting fluid causes a browning of the tissue after two weeks but no alteration in fine structure. If the tissue is allowed to warm up in substituting fluid containing osmium tetroxide, as was done by Van Harreveld and co-workers (*130, 131*), then, of course, the osmium acts as a fixative, and a picture with positive membranes rather similar to osmium fixed material is obtained.

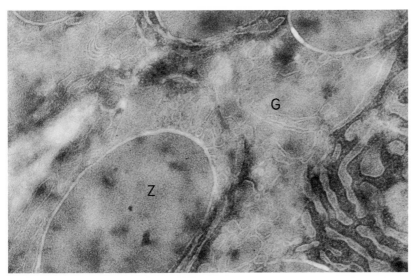

Fig. 4. Portion of cytoplasm of exocrine cell of mouse pancreas prepared as in Fig. 3, except that the section was stained with uranyl acetate. Zymogen granules (Z) can be seen in close association with an extensive and well preserved Golgi region, (G), the membranes of which appear in negative contrast. 60,000×.

Until recently, it seemed that it was impossible to obtain good ice crystal free preparations without the use of some protective agent. However, work by this author and by Van Harreveld, which will be described later, has shown that sufficiently high freezing rates can be obtained to give good preservation. In the author's work *(83)*, small samples were obtained from a dog heart beating *in vitro* by shooting through it a section of small gauge hypodermic needle from a modified rifle. The tissue was caught and frozen rapidly in liquid propane at its melting point; thereafter, it was substituted and embedded. Although there is some ice crystal damage, large parts of cells show good preservation (Fig. 5). Extracellular ice crystals always appear and it seems that the freezing rate must be great enough to prevent intracellular, but not extracellular, ice crystallization. It is estimated that with the velocities used, the specimen passed from the heart into the propane in less than one millisecond and it does not appear that the tissue could dry sufficiently in this time to be protected.

FIG. 5. Cardiac muscle of dog prepared by high speed freezing technique without glycerol protection (see text), freeze-substituted in ethanol at −75°C, em-

Van Harreveld and his colleagues (*130, 131*) use a method of rapid freezing that has been tried by others, namely bringing the tissue into contact with a piece of metal held at low temperature, thereby taking advantage of the high thermal capacity and conductivity of the metal compared with a liquid coolant. An innovation which they make, and which seems to have led to their success, is the mounting of the tissue on a plunger and bringing it into rapid contact with the metal, so much so that it is squashed and frozen at the same time. The details of the method are given in the first paper (*130*). They cool a block of polished silver in liquid nitrogen and continuously flush its surface with dry helium gas at the same temperature to prevent condensation of water or oxygen. The tissue on the plunger hits the metal, is frozen, and then easily detached from the smooth surface. The tissue is then substituted in dry acetone with the addition of 2% OsO_4 at $-85°C$ for three days, allowed to warm up to room temperature, and thereby, fixed. Embedding is in Maraglass. Preservation is good to a depth of 10 μ below the surface which made contact with the metal. Below this, there is ice crystal damage. They have used the method to try to come to a conclusion about the true volume of the extracellular fluid in nervous tissue (*131*). Mouse cerebellum frozen within 30 seconds of circulatory arrest was compared with cerebellum frozen after 8 minutes asphyxiation. Electron micrographs of tissue frozen shortly after circulatory arrest reveal the presence of an appreciable extracellular space between axons of granular layer cells. In asphyxiated tissue, the extracellular space between the axons is either completely obliterated or reduced to narrow clefts between apposing cell surfaces. This is taken to indicate that the *in vivo* extracellular space is more nearly that indicated by electrophysiological methods than by electron micrographs of tissue fixed in a conventional chemical way. Although there are some objections to this work, such as the possibility that surface drying may have taken place leading

---

bedded in Epon, and on-section stained with uranyl acetate in 50% acetone. The fine structural organization of the muscle is well preserved, the myofilaments and Z and H lines being well delineated. The cristae of the mitochondria (M) appear in extremely regular array. Dense glycogen particles are seen around the mitochondria. There is some slight intracellular ice crystal disruption and extensive extracellular ice. It is interesting to see that ice crystals have propagated along the invaginations of the cell surface to the Z line level. N-nucleus. 21,000×. [This micrograph appears in a paper by Monroe *et al.* (*83*).]

to better preservation, and that the tissue may be squashed by the plunger, there is no escaping the fact that a morphological difference is demonstrated between nervous tissue in two different physiological states. The work is a significant application of the technique of freeze-substitution and is also valuable in that measurements of water and NaCl solubilities in acetone at the substitution temperatures were made (see above).

Van Harreveld's group have also turned their attention to the structure of mitochondria after rapid freezing and substitution of fresh tissue (70–72). They show pictures similar to that in Fig. 3 and come to the same conclusion as Bullivant (14) namely that the space within the cristae is closed and the two membranes are fused back-to-back in the same manner as a tight junction. However, evidence from freeze-fracturing (19) and etching (84) is in favor of there being an open space within the cristae.

As mentioned earlier, freeze-substitution is much easier to do and does not seem to have any defects compared with freeze-drying. The earlier work of the author (12) was done very simply, with the exception of the use of helium II which is now known not to be necessary. The substitution was carried out in test tubes immersed in large Dewar flasks containing a dry ice–acetone mixture at −75°C, it was only necessary to add more dry ice every other day. It is convenient, but not essential to have electric deep freezers. Hanzon and Hermodsson (53) analysed the causes of artifacts in freeze-dried material and came to the conclusion that the surface tension forces at impregnation were the most important. These are, of course, largely eliminated by freeze-substitution. These authors believed that they eliminated the freezing itself as a source of artifact because when they did the test of thawing the tissue in osmium solution they obtained good ice free preparations. However, Baker (1) showed that when tissue was thawed in fixative, there was tissue reconstitution leading to a good fine structural picture even though there had been extensive ice crystal formation. This observation led Sjöstrand and Elfvin (115) to criticize freeze-substitution on the grounds that the ice was never substituted out but that it melted when the temperature of the specimen was finally raised to room temperature, and tissue reconstitution gave the appearance of good preservation. This criticism does not stand up for two reasons. First of all, measurements have shown that ice is removed (*28, 32*) [see also Rebhun in discussion after his paper (*102*)]. Van Harreveld *et al.* (*131*) showed that after two days substitution in acetone at —85°C a layer of cells 50–100 μ thick could be rubbed off and presumably this volume had been substituted. Extrapolation from this shows that small cubes of the order of 1 mm on edge could be substituted in two weeks. Second, in freeze-substituted specimens, ice crystal ghosts can be seen (Fig. 1) sometimes coexisting with well preserved areas (Fig. 5). It is obvious in these cases that freeze-substitution has not led to tissue reconstitution.

Recently Pease has done a considerable amount of work involving freeze-substitution of various animal tissues. Only the two papers concerning the technical details will be referred to here (*96, 97*). Excellent structural preservation was obtained by substituting in a eutectic mixture (70%) of ethylene glycol held at —50°C. Freezing of either fresh or glycerol-protected tissue was done by immersion in the supercooled eutectic mixture at —77°C, or by various other methods. A disturbing fact was that whatever the method of freezing and irrespective of whether a protective agent was used, the preservation was uniformly good. Spaces referred to as ice crystals bore no resemblance to those shown in Figs. 1 and 5. Substitution was considered to be completed after a few hours when the tissue turned transparent. Other workers have used periods of days or weeks. It seems likely that the tissue was allowed to warm before substitution was complete, any ice melted, and reconstitution took place. This conclusion is supported by the fact that the well preserved layer of cells were in a subsurface rather than a surface layer. Van Harreveld's group found the surface layer to be well preserved when they froze fresh tissue by contact with cold metal (*130, 131*). Despite these criticisms it is evident that Pease had produced good preservation identical to that of other workers using freeze-substitution.

3. *Direct Sectioning*

As indicated earlier, an ideal solution would be the cutting of thin sections of tissue embedded in the ice in which it was frozen. Fernández-Morán (*29*) did some early work along these lines and obtained some promising results. He froze either fresh or fixed tissue on the stage of a Spencer rotary microtome and cut thin sections onto saline or glycerol solutions using thermal advance with the pointer set on zero. The illustrations he presented were remarkably good considering that conventional methods of thin sectioning were also in their infancy at the time. Bernhard and Nancy (3) have cut frozen sections of glutaraldehyde-fixed tissue which was infiltrated with low concentrations of gelatin before freezing. Their technique, as it stands at present, is described in a recent paper (2). Tissue is fixed in glutaraldehyde, embedded in 20% gelatin, dehydrated in 50% glycerol, frozen in liquid nitrogen, and sectioned at −35°C onto 40% dimethyl sulfoxide using an ultramicrotome in a deepfreeze. The sections are stained with a variety of heavy metal salts and show membranes in negative contrast similar to their appearance after freeze-drying or freeze-substitution. The reason for developing the technique was to produce sections on which ultrastructural cyto-chemistry could be carried out (64). They reasoned that as freeze sectioning involves no embedding in the conventional sense, there would be a reduced possibility of inactivation of the enzymes be-cause of reaction with the embedding medium. This reasoning is borne out by the good results they obtained.

With the same goal as Bernhard's group, that is to produce sections for ultrastructural cytochemistry that have not been fixed or exposed to solvents, Christensen (22) recently produced extremely encourag-ing results, when cutting thin sections of rapidly frozen fresh tissue. This is nearer to the ideal than Bernhard's work in that no fixative or other treatment is used. Small pieces of tissue are frozen rapidly against a block of copper at liquid nitrogen temperature, a modifica-tion of Van Harreveld's technique which was mentioned earlier. An insulated extension from the specimen arm of a conventional ultra-microtome carries the frozen specimen at the level of a diamond knife mounted in the bottom of a Dewar flask containing either liquid nitrogen or Freon 22. Sections of about 1000 Å in thickness are cut dry, mounted on grids, and allowed to come to room tempera-ture as they are dried in a jet of nitrogen gas. Figure 6 shows a specimen prepared in this manner. It should be emphasized that this section is unfixed, unembedded, unstained, and has never come in contact with any extraneous solvent. Many of the features of con-ventional thin sections can be seen.

Both Reichert and LKB have produced attachments that enable

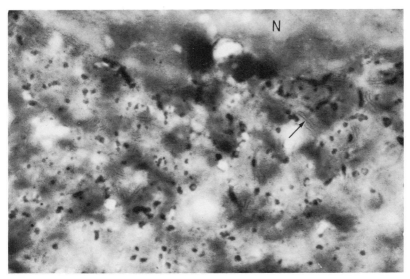

Fig. 6. Nucleus (N) and cytoplasm of rat hepatocyte prepared by Christensen's frozen thin section method (see text). The material in this section is unfixed, unembedded, and unstained. Profiles of rough endoplasmic reticulum can be recognized (arrow) by virtue of the density of the ribosomes. 15,000×. (Micrograph provided by courtesy of Dr. A. K. Christensen.)

their own ultramicrotomes to be used to cut thin sections of frozen tissue. Reference should be made to these companies for details.

C. Freeze-Etching and Fracturing

Freeze-etching and fracturing are presently the techniques which are being most used in electron microscopy involving freezing as a preparation method. What follows is a description of the methods being used with some discussion of the interpretation of results. However, references on applications are quoted on a selective rather than an exhaustive basis.

The principle of freeze-etching is that a small amount of the surface ice of a frozen specimen is allowed to sublime off in a vacuum, leaving nonvolatile structures as projections which can be replicated. The basis for this approach was provided by Hall (*51*) and independently by Meryman (77) who both produced replicas of the surfaces of frozen aqueous solutions. Meryman and Kafig (82) made an apparatus in which they were able to fracture frozen material and etch and replicate the resultant surface. Although their results with biological materials were disappointing, they must certainly be credited with an apparatus which is the forerunner of the machines currently in use. Also of importance to freeze-etching and fracturing, they demonstrated recrystallization of amorphous ice on warming briefly to —96°C and suggested that such recrystallization could be important to —130°C. The first successful biological application was the work of Steere (123). He froze suspensions of material, usually virus crystals, in drops, cut slices off the top of the drop with a scalpel, and then allowed the preparation to warm up to about —90°C in a vacuum, at which point the water vapor in the air, which had condensed on the surface, and also some of the surface ice evaporated. He then made a heavy metal, carbon backed replica of this surface, floated it off, and dissolved any attached biological material before examination in the electron microscope. Haggis (50) made an ingenious apparatus to fracture frozen blood smears in a vacuum and replicate the fracture surface. Electron micrographs showed the outlines of cells, and it was claimed that the hemoglobin molecules in the cytoplasm could be recognized. A method of freeze-etching that has given some remarkable results in that devised by Moor et al. (87). The apparatus is quite complicated and reference should be made to the original paper for details. Essentially it consists of an ultramicrotome with the facility for cutting thin sections at low temperature, mounted in a vacuum evaporator. Tissue is usually glycerinated to avoid ice crystal formation, frozen in Freon, and mounted on the cold specimen stage of the microstome. High vacuum is obtained and the temperature of the specimen raised to —100°C and held constant there by alternate cooling and heating under control of an electronic feedback system governed by a thermistor in the stage. Thin sections are cut with a liquid nitrogen cooled steel knife. The heel of the knife is then placed over the specimen to act as a cold trap and the remaining freshly cut surface is left for a short while (usually on the order of a minute) for a few hundred Å thickness of the ice to sublime. This leaves the nonaqueous components of the layer standing above the ice. A carbon platinum replica of the surface is made and backed with carbon. The specimen is taken out of the vacuum, the replica floated off, and the adhering biological material dissolved before observation in the electron microscope. The remarkably good preservation and detail obtained by the use of this technique can be seen in Figs. 7 and 8. The cutting is more in the nature of cleaving than microtomy in the usual sense. The cleavage plane goes along the surface of organelles in some cases and cross fractures others, so that the replica contains much more three dimensional information than the usual thin section. Apart from this, the other great advantage of the method is that once the tissue is frozen, its temperature never rises above −100°C until after the replica is made. Moor and Mühlethaler (*88*) froze yeast cells rapidly without glycerination and demonstrated appreciable retention of viability after thawing. Similar cells were examined by freeze-etching. Branton and Moor (*9*) grew onions in glycerol solutions and looked at the roots by freeze-etching. In both these cases the investigators were looking at replicas of the freeze-fractured and etched inner surfaces

Fig. 7. Junction between two cells in small intestine of mouse prepared in Moor freeze-etching machine. Etching reveals membranes in relief, the unit membrane structure being visible (arrow). The cytoplasm has a fine granular appearance caused by the etching. The encircled arrow indicates the direction of shadowing. 60,000×. (Micrograph provided by courtesy by Dr. L. A. Staehelin.)

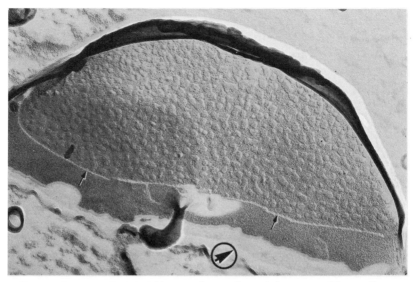

Fig. 8. Freeze-etched lecithin membrane formed in water. The replica was prepared in a Moor machine. The fracture plane leaves a number of plaques on the surface (see text). Deep etching shows a ridge (arrows), below which is revealed a smooth surface. The encircled arrow indicates the direction of shadowing. 80,000×. (Micrograph provided by courtesy of Dr. L. A. Staehelin.)

of cells that had only been changed from their living state by the single step of rapid freezing. There is a disadvantage of the technique: A replica rather than a thin section is produced and it contains only morphological information. A number of studies have been carried out using the machine (*4, 6–11, 23, 24, 26, 57–60, 84–90, 100, 104–106, 118–122*), and the structure of cells and organelles as seen in thin sections has been largely confirmed (see Figs. 7 and 8). The most interesting new findings have concerned the structure of membranes and these will be discussed later. The freeze-etch equipment after Moor's design is manufactured by the Balzers Co., Liechtenstein, and is obtainable in the United States from Bendix-Balzers Vacuum Inc., Rochester, New York.

The potential of the technique on the one hand and the complexity and expense of the Moor machine on the other has led others to look for simpler ways of doing all or part of what it accomplishes. Koehler

(*58*) has made a device essentially the same as the Moor machine. Reduction in expense was achieved by using an ordinary vacuum evaporator as the basis and controlling the specimen temperature manually by operation of the valve admitting liquid nitrogen into the specimen stage.

Steere has made a modular freeze etching device which can be mounted on any vacuum evaporator. The tissue is cut inside of the vacuum with a cooled scalpel blade attached to a feed through arm, which can be manipulated from outside. The apparatus is described in detail in a recent paper (*124*).

Bullivant and Ames (*19*) produced a simple-freeze-fracture device which has come to be known as a type I apparatus. The frozen tissue was cut under liquid nitrogen with a cold razor blade outside of the evaporator rather than inside with a cooled microtome. It was subsequently covered with a cold metal lid to prevent contamination and transferred to the evaporator under the surface of the nitrogen. When a good vacuum was obtained, the protective lid was lifted and a carbon platinum replica made. This was digested off and examined in the electron microscope. This device is simple and only costs about $50 to make, over and above the price of a standard vacuum evaporator. Compared with the Moor apparatus, it does not etch and will not precisely fracture very small specimens. However, it does produce comparable replicas.

The majority of the recent work of the present author has been done with a type II freeze-fracture apparatus. This was referred to in a footnote to the original paper (*19*) and has also been subsequently described (*20*). A diagram of the three parts of the block, together with their dimensions, is shown in Fig. 9. All parts are turned from 3 inch diameter brass cylinder. The lower part (5) is screwed into the bottom of a vessel made from the lower half of a 400 ml polyethylene beaker, which is itself attached to a lucite base (Fig. 10C). The middle part (3), with the two shadowing tunnels, locks in a particular position with respect to the lower part by means of a locating pin. The platinum–carbon and carbon electrodes are prealigned so that when the lucite base is pushed between the electrode uprights and against a stop, the shadowing tunnels are correctly aligned.

The use of the apparatus will be described with reference to Figs.

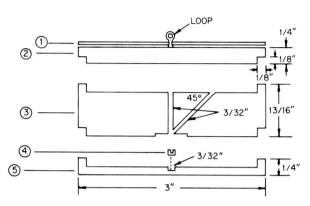

FIG. 9. Diagram of type II freeze-fracturing device, giving dimensions in inches.

10 and 11. Small pieces of tissue, either fixed with glutaraldehyde or fresh, are soaked in 25% glycerol in buffer in the refrigerator for at least 30 minutes. A piece of tissue is inserted in the small brass specimen holder [(4) in Fig. 9] and is rapidly frozen by dropping it into liquid Freon 12 (dichlorodifluoromethane) at its melting point of −155°C. The other parts of the apparatus are already cooled to liquid nitrogen temperature under the surface of the nitrogen in a large styrofoam container (Fig. 10A). A pair of forceps are cooled by dipping in liquid nitrogen and used to transfer the specimen from the Freon through the nitrogen surface and position it in the hole in the center of the lower block, leaving a part of the frozen tissue protruding above the level surface of the block (Fig. 10A). A clean, new razor blade held in a pair of hemostats is cooled in the nitrogen and used to cleave the specimen with a single sweep (Fig. 10B). The middle block and the lid are quickly put in position (Fig. 10C); the container, brimming to the edge with liquid nitrogen, is positioned beneath the electrodes in the evaporator (Fig. 11A). A hook from the small electric crane is attached to the loop on the lid. The bell jar is put in place and pumping is commenced with the rotary pump. The liquid nitrogen boils rapidly and then freezes under the reduced pressure. After a period of about ten minutes, the nitrogen sublimes away and the pressure becomes low enough to turn over to the oil diffusion pump. Once a pressure of 10^{-4} mm Hg is obtained, the electrodes are outgassed by heating them to a cherry red color. With the evaporator used in this laboratory (Kinney model KSE-2, with nitro-

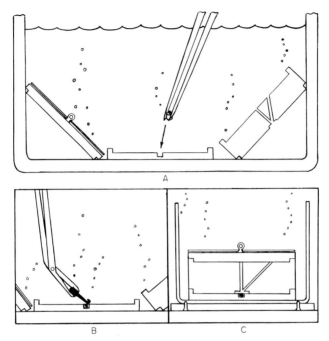

Fɪɢ. 10. Diagram of initial stages of use of type II freeze-fracture device under liquid nitrogen. (A) Insertion of prefrozen specimen. (B) Cleaving of specimen. (C) Assembled device ready for transfer to evaporator. For simplicity, the lower part is only shown in the polyethylene container and fixed to the lucite base in C. In actual fact it is always set up in this way.

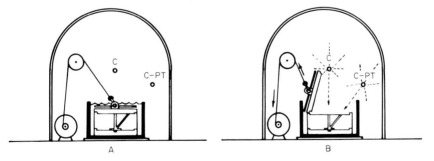

Fɪɢ. 11. Diagram showing later stages of use of type II freeze-fracture device. (A) Container with device covered with liquid nitrogen located in evaporator and lid connected to crane. (B) After vacuum is established, the lid is lifted and shadowing with carbon–platinum and backing with carbon is done.

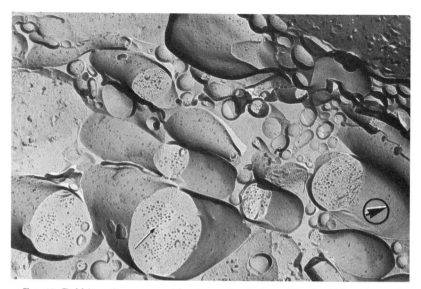

Fig. 12. Rabbit retina prepared by freeze-fracturing without etching. A number of unmyelinated axons can be seen. They fracture at right angles to their long axes; broken-off neurotubules can be seen protruding above these fracture faces (arrow). The encircled arrow indicates the direction of shadowing. 30,000✕.

gen trap), a pressure of 10^{-5} mm Hg is obtained within 30 minutes of the start of the experiment. At this pressure, the lid is lifted by the electric crane and the specimen is shadowed with carbon–platinum and backed with evaporated carbon (Fig. 11B). The best method of evaporating carbon–platinum is to wind about 1 inch of 0.004 inch Pt wire onto the junction between two $\frac{1}{32}$-inch diameter carbon points and heat to white heat by passing current. The electrodes are positioned about 8 cm from the specimen. The amount of shadow can be monitored by watching the polished upper surface of the middle block. A deep purple color indicates sufficient shadow. At the time of shadowing, the temperature of the specimen is certainly below $-150°$C. It has recently been found that a thin metal container can be substituted for the polyethylene one and gives equally good results. With a metal container the specimen temperature rises more rapidly but the nitrogen evaporates faster; hence, operating vacuum is obtained earlier. In this case shadowing is done at about $-140°$C.

Temperatures are monitored by inserting a copper constantin thermocouple in a hole in the lower block and taking the signal produced via a feed-through to a DC operational amplifier. The temperatures at which replication is done are generally below the range within which recrystallization of the amorphous frozen glycerol–ice phase would take place (*69*). Depending on the type of specimen, sulfuric acid, nitric acid, strong household bleach, or enzyme preparations may be used to digest the tissue off the replica. The replicas are washed with distilled water, mounted on grids, and examined in the electron microscope. Examples of micrographs produced by the freeze-fracture technique are shown in Figs. 12–15. In the original type I apparatus, the cleaved specimen was protected from condensation of contamination by the lid and the covering of liquid nitrogen. When the lid was lifted, the specimen was exposed to contaminant

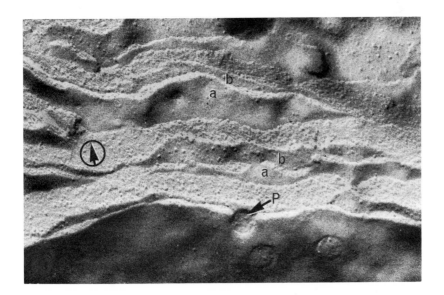

FIG. 13. Nuclear membrane and adjacent cytoplasm of mouse exocrine pancreas cell prepared by freeze-fracturing. A nuclear pore (P) is viewed from within the nucleus looking out. The fracture breaks across the endoplasmic reticulum revealing alternately smooth (a) and particle covered faces (b). The particle covered faces are always seen when looking from within a cisternal space. The encircled arrow indicates the direction of shadowing. 100,000✕.

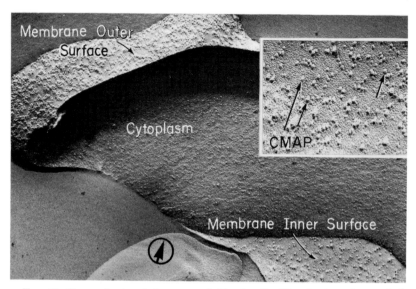

FIG. 14. Freeze-fractured human red blood cell. The fracture faces of the cell membrane are associated with a small particulate component which is more abundant on the outer than the inner surface aspects. 42,0000×. *Insert:* Higher magnification of the inner aspect of the plasma membrane shows that the cell membrane associated particles (CMAP, arrows) cover a small percentage of the total surface area and tend to be arranged in small clusters and chains. 77,000×. The encircled arrow indicates the direction of shadowing in both cases (*138a*). (Micrographs provided by courtesy of Dr. R. S. Weinstein.)

molecules arriving from all directions and some of the granularity seen in the specimens was attributed to contamination deposited in this fashion. In the type II apparatus, the specimen is only exposed to the contamination arriving in the small solid angle represented by the tunnels. In the type I apparatus, frost was picked up by the outside of the polyethylene container during transfer. Once the nitrogen had gone, the container warmed up more rapidly than the block and the frost sublimed. However, some condensation on the metal lid remained as a source of contaminant water. In the type II apparatus an additional plastic lid covers the metal lid [(1) in Fig. 9]; any frost picked up on this sublimes as the plastic warms up before the lid is lifted. Figure 13 shows fracture faces of the endoplasmic reticulum produced by a type II apparatus. These are

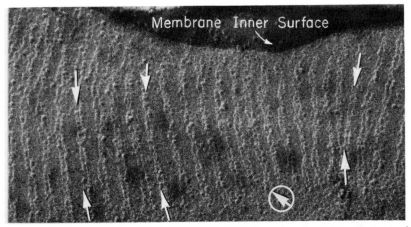

Membrane Inner Surface

FIG. 15. Freeze-fracture of sheep red blood cell passing from inner aspect of the membrane through the cytoplasm. The parallel row of 100 Å particles may reflect paracrystalline packing of hemoglobin molecules. The encircled arrow indicates the direction of shadowing. 95,000× (*134a*). (Micrograph provided by courtesy of Dr. R. S. Weinstein.)

alternately rough and smooth. The same replica made with a type I apparatus shows particles on all faces, indicating that some of these are contaminants and that the use of the type II apparatus does reduce contamination.

Freeze-fracturing produces views of membranes similar to those seen after etching (Figs. 12 and 13). Etching reveals cross-fractured membranes in greater relief (Fig. 7). On the other hand, it may mask certain structures. Etching leaves a granular surface where ice has sublimed away; granular or fibrillar structures may disappear among this granularity. Examples are the structure of the chromatin of the nucleus and the basement membrane region of cells both of which are seen after freeze-fracturing but lost when the specimen is etched. A number of studies have been made using simple freeze-fracturing devices (*17–19, 61–63, 76, 117, 133–139*).

There have been a number of attempts to modify either the type I or the type II apparatus to permit etching. Bullivant (*16*) arranged a type I apparatus so that the specimen could be heated relative to the lid. Etching was produced, but this was not entirely reproduceable owing to the difficulty of holding the temperature in the −100°C range during etching. This is critical, for a 5°C variation on either side results either in no etching or in excessive etching accompanied by recrystallization. The only way to achieve the desired result is to heat the specimen and then turn off the current at such a point that the specimen temperature would level onto a plateau at the −100°C mark. In the Moor apparatus, the temperature of the specimen mount is stabilized at −100°C before the fracture is made. A similar modification of the type I apparatus has been made by Weinstein and Someda (*135, 137*). This produces very good etching in the case of red cells (*131a*). McAlear and Kreutziger (*63, 75*) made a modification of the type II apparatus. In this, the specimen was fractured in the vacuum by a razor blade set in the part of the block containing the tunnels. There was no upper lid. Rotation of the upper block via a mechanical feed-through caused the razor blade to pass through the specimen. The rotation was halted by a stop so that the tunnels finished up aligned with the specimen and the electrodes. To give etching, the specimen was warmed either by thermal contact of the specimen holder with the warmer baseplate of the evaporator, or by radiant heat through one of the tunnels. Unfortunately, the temperature cannot be controlled as accurately as in the Moor apparatus, but sufficiently good results have been obtained to make it appear that some further modification will give comparable etching.

The interpretation of freeze-etched and freeze-fractured replicas is difficult. One difficulty is in deciding where the fracture plane goes in a membrane; the other concerns the nature of the many small particles seen on membranes.

It was initially thought that the break would go along the surface of a membrane but this view was challenged by Branton (*7*) using comparisons between etched and nonetched material and also demonstrating a ridge at the base of fractured membranes that he interpreted as being the edge of a half membrane sheet. A similar edge is shown in Fig. 8. Weinstein (*136*) believes that this half membrane may be a "pseudo membrane" deposited from material left behind during sublimation. Branton put forward the theory that fractures proceed along the inner hydrophobic surfaces of membranes. He also looked at artificial bimolecular lipid layers (*24*); by radioactively labeling one half, he was able to show that, when frozen and fractured, it broke down the middle. He has applied freeze-etching to various membrane systems such as the chloroplast (*10*), the retina (*23*), and the myelin sheath (*7*). An extensive discussion of membrane structure is beyond the scope of this chapter and can be found in a recent review by Branton (*8*). His main conclusions can, however, be stated.

In membrane systems such as those of the chloroplast, which are weakly birefringent, freeze-etching shows many small particles. Such membranes are composed of repeating lipoprotein subunits. Membranes such as those of the myelin sheath, which are strongly birefringent, show smooth surfaces on freeze-etching and have a structure corresponding to the "unit membrane" of Robertson. Mühlethaler *et al.* (*90*) saw the same granular structure in freeze etched chloroplasts but interpreted it as being on the outside of the membrane. Staehelin (*120*) has freeze-etched artifical lipid bilayers and has shown a smooth surface covered with plateaus of half membrane thickness (Fig. 8). He interprets this to mean that the fracture plane skips back and forth from the center to the outside of the membrane. However, he has recently shown that the appearance of membranes depends on the temperature of fracturing and replicating (personal communication). The plateaus are not seen in material fractured and replicated below $-140°C$; this author has never observed them in freeze-fractured material. Bullivant and Weinstein (*21*) postfixed and embedded tissue which had been prefixed in glutaraldehyde, glycerinated, and freeze-fractured. Sections at right angles to the fracture plane showed either intact unit membranes or no membranes. This was interpreted to mean that the fracture plane went along the outer surface of the membrane. More recent experiments by Bullivant (*17*), in which the fractured surface was protected by a thin coat of evaporated carbon before thawing and postfixing, led to an equivocal answer; lines having the appearance of a half membrane were seen under the carbon in some instances. At present there appears to be most evidence to back up Branton's hypothesis that the break is within the membrane and that the particles seen are within it also. An unequivocal answer could be obtained by looking at replicas obtained by shadowing both sides of the fracture plane along a particular membrane. Several laboratories are attempting to do this at present.

The other difficulty of interpretation is the nature of particles on membrane fracture faces and in other places (Figs. 13 and 14). Originally, Bullivant and Ames (*19*) suggested that particles were contaminant artifacts in many cases because the number of particles was reduced each time an improvement expected to reduce contamination was made in the apparatus. As mentioned earlier, the use of the type II apparatus gives a great reduction in contamination, and yet particles still appear on some faces while others are relatively smooth (Fig. 13). Presumably these remaining particles are not artifacts. Arguments against the particles seen in freeze-etch replicas being artifacts were produced by Branton and Park (*10*). Weinstein *et al.* (*139*) also have evidence that the majority of particles are not contamination artifacts, based on the fact that membranes of artificial myelin figures appear smooth. The particles can be correlated with quantasomes in chloroplasts (*10*), with hemoglobin molecules in red blood cells (*134*) (Fig. 15), and possibly with particles seen in nexuses by other methods (*17, 18, 62, 76*). This gives more substance to their reality in these cases, but there is still need for caution in their interpretation (*117*).

If the particles are real, then there is another difficulty. Figure 13 serves as a good example to base a discussion of this point on. It shows faces produced by a stepwise fracture going across the endoplasmic reticulum of a cell in the exocrine pancreas. Alternate fracture faces are covered with 100 Å particles (face a) and then relatively particle free (face b). It was originally thought that the particles were ribosomes, but their smaller size and the topology of the situation, which could be worked out by counting off alternate cytoplasmic and cisternal spaces starting at the nuclear membrane, led to the conclusion that the particles were present in the center of a split membrane. The smooth membrane must be the three dimensional mirror image of the particle covered one and yet it contains no depressions out of which the particles could have been pulled during the fracturing process. This excess of particles over depressions is general in freeze-etched and fractured replicas, although there are some instances where a few depressions can be seen (*10, 17, 120*). The depressions are presumably lost as an artifact of the method. One early explanation with the Moor machine was that the cutting of a thin section off the surface of the specimen was such an asymmetric process that particle faces were always left and depression faces always pulled away.

In freeze-fracturing, as developed by Bullivant and Ames, the specimen is broken approximately into two parts and the explanation of asymmetry cannot hold. It seems more likely that some plastic deformation during fracture (*4, 23*), an effect of the etching, the localized heating during shadowing, or contamination, accounts for the loss of the depressions (*17*). Depressions in membranes are difficult to see, except under the best conditions. Even the array of depressions in the "close" junction may be obscured by contamination (*61*). In this laboratory we have also found that this array can be covered with what is apparently contamination, although the quality of shadowing also plays a role in its visualization.

The appearance of two distinct faces of membranes, as described earlier for the endoplasmic reticulum in the pancreas, seems to be a general phenomenon (*8, 17*). The plasma membrane *outer* aspect and the *inner* aspect of membrane-bound cytoplasmic organelles such as mitochondria or the endoplasmic reticulum appear to be covered with many 100–200 Å particles. The inner aspect of the plasma membrane and the outer aspect of cytoplasmic organelles are relatively smooth and may have a few depressions. This asymmetry of particle population was first reported explicitly by Someda and Weinstein (*117*). It is interesting that this asymmetry is correctly maintained when the plasma membrane is turned inside out to give a cytoplasmic organelle, as is assumed to occur during invagination and pinching off (*8, 17*).

Freeze-etching and fracturing techniques provide face-on views of membranes and it is natural that they have found a use in the study of junctions between cells. The tight junctions in epithelia have been described by Kreutziger (*62*) and Staehelin *et al.* (*122*). In such a junction, the plasma membrane outer aspect shows a series of interconnecting concertina-like ridges. The inner aspect shows similarly arranged furrows. These authors assumed that they were seeing the outer surface of the membrane, and hence placed the ridges in the intercellular space within the junction. The nexus or "gap" junction, originally shown to have a hexagonal array in face view after lanthanum staining (*107*) has been looked at using the Bullivant and Ames type II device (*17, 18, 76*) or a modification of it (*61, 62*). The outer aspect of the plasma membrane shows a close packed array of particles in the junctional area, and the inner aspect a close packed (90 Å

center-to-center) array of depressions. Workers who believe that
fracturing reveals the true surface of membranes have naturally
placed the particles in the gap between the membranes and the depressions on the inside surface of the plasma membrane (*62, 76*). It
seems more likely that the arrays of particles and depressions are the
complementary three-dimensional mirror images of one another produced by fracturing within membranes. Again, absolute proof of
this will have to await the examination of both sides of a particular
break within such a junction. The ridges in tight junctions, the arrays
of particles in "gap" junctions and the majority of randomly distributed particles are all seen on a particular side of membrane
breaks. Conversely, the furrows, arrays of depressions, and the few
depressions seen in nonjunctional areas are on the other side. There
are, thus, two distinct complementary faces; this fits in best with
Branton's idea of a single split within membranes. If the break were
capable of revealing either outside face of a membrane, then there
should be four complementary faces seen (*8, 17, 23*).

Freeze-etching and freeze-fracturing certainly provide a new and
interesting view of cells and great progress in techniques has been
made recently. Before full value can be obtained from the use of the
techniques, there will have to be corresponding understanding of the
events taking place when cells are frozen, fractured, etched, and
replicated.

Apart from the information in this chapter there are reviews of
freeze-etching and freeze-fracturing by Moor (*86*), Weinstein and
Someda (*137*), and Koehler (*60*) which should be consulted.

IV. Conclusions

Freezing methods are an approach to electron microscopy in which
some of the artifact producing steps of conventional techniques are
eliminated. In particular, some of the techniques offer a chance to
look at the cell without the removal of water and in a state where
it is potentially viable. Freeze-drying and substitution, and to a
greater extent the thin sectioning of tissue rapidly frozen and embedded
only in its own ice, will become increasingly useful in the preserva-

tion of the location of ions and molecules for subsequent cytological location by chemical and physical methods. Freeze-etching and fracturing techniques should come to the forefront in the investigation of membrane substructure and also possibly in work on structures such as chromosomes which are particularly labile to conventional techniques. Most of the freezing methods mentioned in this review have only been developed for electron microscopy in the past ten years; it is certain that in the very near future they will occupy a large segment of the methodology of electron microscopy.

ACKNOWLEDGMENTS

The author's work described in this chapter has been supported by the National Institutes of Health, U.S. Public Health Service, Bethesda, Maryland. The micrographs were kindly provided by Dr. A. Kent Christensen, Dr. L. Andrew Staehelin, and Dr. Ronald S. Weinstein. The author is also grateful to Dr. Staehelin and Dr. Weinstein for their helpful discussions and advice. Similarly, he is indebted to Mr. J. J. Fields for technical assistance and preparing the illustrations.

REFERENCES

1. Baker, R. F. Freeze thawing as a preparatory technique for electron microscopy. *J. Ultrastruct. Res.* **7**, 173 (1962).
2. Bernhard, W., and Leduc, E. H. Ultrathin frozen sections. 1. Methods and ultrastructural preservation. *J. Cell Biol.* **34**, 757 (1967).
3. Bernhard, W., and Nancy, M-T. Coupes a congélation ultrafines de tissue inclus dans la gélatine. *J. Microscopie* **3**, 579 (1964).
4. Bertaud, W. S., Rayns, D. G., and Simpson, F. O. Myofilaments in frozen-etched muscle. *Nature* **220**, 381 (1968).
5. Bondareff, W. Morphology of particulate glycogen in guinea pig liver revealed by electron microscopy after freezing and drying and selective staining *en bloc. Anat. Record* **129**, 97 (1957).
6. Branton, D. Fracture faces of frozen membranes. *Proc. Natl. Acad. Sci. U.S.* **55**, 1048 (1966).
7. Branton, D. Fracture faces of frozen myelin. *Exptl. Cell Res.* **45**, 703 (1967).
8. Branton, D. Membrane structure. *Ann. Rev. Plant Physiol.* **20**, 209 (1969).
9. Branton, D., and Moor, H. Fine structure in freeze-etched *Allium cepa L.* root tips. *J. Ultrastruct. Res.* **11**, 401 (1964).
10. Branton, D., and Park, R. B. Subunits in chloroplast lamellae. *J. Ultrastruct. Res.* **19**, 283 (1967).
11. Branton, D., and Southworth, D. Fracture faces of frozen *Chlorella* and *Saccharomyces* cells. *Exptl. Cell Res.* **47**, 648 (1967).
12. Bullivant, S. The staining of thin sections of mouse pancreas prepared by

the Fernández-Morán helium II freeze substitution method. *J. Biophys. Biochem. Cytol.* **8**, 639 (1960).

13. Bullivant, S. Consideration of membranes and associated structures after cryofixation. *Proc. 5th Intern. Conf. Electron Microscopy, Philadelphia, 1962* Vol. 2, Art. R2 (abstr.). Academic Press, New York, 1962.

14. Bullivant, S. Observations on the structure of biological membranes. *J. Cell Biol.* **23**, 16a (1964).

15. Bullivant, S. Freeze-substitution and supporting techniques. *Lab. Invest.* **14**, 1178 (1965).

16. Bullivant, S. Freeze-fracturing and freeze-etching. *New Zealand Med. J.* **66**, 387 (1967).

17. Bullivant, S. Freezing-fracturing of biological materials. *Micron* **1**, 46 (1969).

18. Bullivant, S. Particles seen in freeze-fracture preparations. *Proc. Electron Microscopy Soc. Am.* p. 206 (1969).

19. Bullivant, S., and Ames, A., III. A simple freeze-fracture replication method for electron microscopy. *J. Cell Biol.* **29**, 435 (1966).

20. Bullivant, S., Weinstein, R. S., and Someda K. The type II simple freeze-cleave device. *J. Cell Biol.* **39**, 19a (1968).

21. Bullivant, S., and Weinstein, R. S. A thin section study of the path of fracture planes along frozen membranes. *Anat. Record* **163**, 296 (1969) (abstr.).

22. Christensen, A. K. A simple way to cut frozen thin sections of tissue at liquid nitrogen temperatures. *Anat. Record* **157**, 227 (1967).

23. Clark, A. W., and Branton, D. Fracture faces in frozen outer segments from the guinea pig retina. *Z. Zellforsch. Mikroskop. Anat.* **91**, 586 (1968).

24. Deamer, D. W., and Branton, D. Fracture planes in an ice bilayer model membrane system. *Science* **158**, 655 (1967).

25. Dupouy, G., Perrier, F., and Durrieu, L. L'observation de la matière vivant au moyen d'un microscope électronique fonctionnant sous très haute tension. *Compt. Rend.* **251**, 2386 (1960).

26. Du Praw, E. J. Configuration of viral chromosomes in solution, as determined by freeze etching. *J. Cell Biol.* **35**, 35a (1967).

27. Elfvin, L-G. The ultrastructure of the plasma membrane and myelin sheath of peripheral nerve fibers after fixation by freeze-drying. *J. Ultrastruct. Res.* **8**, 283 (1963).

28. Feder, N., and Sidman, L. R. Methods and principles of fixation by freeze-substitution. *J. Biophys. Biochem. Cytol.* **4**, 593 (1958).

29. Fernández-Morán, H. Application of the ultrathin freezing sectioning technique to the study of cell structures with the electron microscope. *Arkiv Fysik* **4**, 471 (1952).

30. Fernández-Morán, H. Electron microscopy of nervous tissue. *In* "Metabolism of the Nervous System" (D. Richter, ed.), p. 1. Pergamon Press, Oxford, 1957.

31. Fernández-Morán, H. Electron microscopy of retinal rods in relation to localization of rhodopsin. *Science* **129**, 1284 (1959).
32. Fernández-Morán, H. Low temperature preparation techniques for electron microscopy of biological specimens based on rapid freezing with liquid helium II. *Ann. N.Y. Acad. Sci.* **85**, 689 (1960).
33. Fernández-Morán, H. Direct study of ice crystals and of hydrated systems by low temperature electron microscopy. *J. Appl. Phys.* **31**, 1841 (1960) (abstr.).
34. Fernández-Morán, H. The fine structure of vertebrate and invertebrate photoreceptors as revealed by low temperature electron microscopy. *In* "The Structure of the Eye" (G. K. Smelser, ed.), p. 521. Academic Press, New York, 1961.
35. Fernández-Morán, H. Lamellar systems in myelin and photoreceptors. *In* "Macromolecular Complexes" (M. V. Edds, Jr., ed.), p. 113. Ronald Press, New York, 1961.
36. Fernández-Morán, H. Cell membrane ultrastructure—low temperature electron microscopy and X-ray diffraction studies of lipoprotein components in lamellar systems. *In* "Ultrastructure and Metabolism of the Nervous System" (S. R. Korey, ed.), p. 235. Williams & Wilkins, Baltimore, Maryland, 1962.
37. Fernández-Morán, H. New approaches to the study of biological ultrastructure by high resolution electron microscopy. *Symp. Intern. Soc. Cell Biol.* **1**, 411 (1962).
38. Fernández-Morán, H., and Dahl, A. O. Electron microscopy of ultrathin frozen sections of pollen grains. *Science* **116**, 465 (1952).
39. Finck, H. An electron microscopy study of basophil substances of frozen dried rat liver. *J. Biophys. Biochem. Cytol.* **4**, 291 (1958).
40. Fisher, H. W., and Cooper, T. W. Electron miroscope studies of the microvilli of HeLa cells. *J. Cell Biol.* **34**, 569 (1967).
41. Gersh, I. The Altmann technique for fixation by drying while freezing. *Anat. Record* **53**, 309 (1932).
42. Gersh, I. The preparation of frozen-dried tissue for electron microscopy. *J. Biophys. Biochem. Cytol.* **2**, Suppl., 37 (1956).
43. Gersh, I. Freeze-substitution. *Federation Proc.* **24**, Suppl. 15, S-233 (1965).
44. Gersh, I., and Stephenson, J. L. Freezing and drying of tissues for morphological and histochemical studies. *In* "Biological Applications of Freezing and Drying" (R. J. C. Harris, ed.), p. 329. Academic Press, New York, 1954.
45. Gersh, I., Isenberg, I., Bondareff, W., and Stephenson, J. L. Submicroscopic structure of frozen dried liver specifically stained for electron microscopy. Part II. Biological. *Anat. Record* **128**, 149 (1957).
46. Gersh, I., Isenberg, I., Stephenson, J. L., and Bondareff, W. Submicroscopic structure of frozen-dried liver specifically stained for electron microscopy. Part I. Technical. *Anat. Record* **128**, 91 (1957).
47. Gersh, I., Vergara, J., and Rossi, G. L. Use of anhydrous vapors in post-

fixation and in staining of reactive groups of proteins in frozen-dried specimens for electron microscopic studies. *Anat. Record* **138**, 445 (1960).
48. Glick, D., and Malmtrom, B. G. Simple and efficient free-drying apparatus for the preparation of embedded tissue. *Exptl. Cell Res.* **3**, 125 (1952).
49. Grunbaum, B. W., and Wellings, S. R. Electron microscopy of cytoplasmic structure in frozen dried mouse pancreas. *J. Ultrastruct. Res.* **4**, 73 (1960).
50. Haggis, G. H. Electron microscope replicas from the surface of a fracture through frozen cells. *J. Biophys. Biochem. Cytol.* **9**, 841 (1961).
51. Hall, C. E. A low temperature replica method for electron microscopy. *J. Appl. Phys.* **21**, 61 (1950).
52. Hallet, J. Reply to question in Stowell (*129*, p. S-43).
53. Hanzon, V., and Hermodsson, L. H. Freeze-drying of tissues for light and electron microscopy. *J. Ultrastruct. Res.* **4**, 332 (1960).
54. Hanzon, V., Hermodsson, L. H., and Toschi, G. Ultrastructural organization of cytoplasmic nucleoprotein in the exocrine pancreas cells. *J. Ultrastruct. Res.* **3**, 216 (1959).
55. Harris, R. J. C., ed., "Freezing and Drying." Blackwell, Oxford, 1951.
56. Harris, R. J. C., ed., "Biological Applications of Freezing and Drying." Academic Press, New York, 1954.
57. Jost, M. Die Ultrastruktur von *Oscillatoria rubescens* D.C. *Arch. Mikrobiol.* **50**, 211 (1965).
58. Koehler, J. K. Fine structure observations in frozen-etched bovine spermatozoa. *J. Ultrastruct. Res.* **16**, 359 (1966).
59. Koehler, J. K. Freeze-etching observations on nucleated erythrocytes with special reference to the nuclear and plasma membranes. *Z. Zellforsch. Mikroskop. Anat.* **85**, 1 (1968).
60. Koehler, J. K. The technique and application of freeze-etching in ultrastructure research. *Advan. Biol. Med. Phys.* **12**, 1 (1968).
61. Kreutziger, G. O. Specimen surface contamination and the loss of structural detail in freeze-fracture and freeze-etch preparations. *Proc. Electron Microscopy Soc. Am.* p. 138 (1968).
62. Kreutziger, G. O. Freeze-etching of intercellular junctions of mouse liver. *Proc. Electron Microscopy Soc. Am.* p. 234 (1968).
63. Kreutziger, G. O., and McAlear, J. H. Three dimensional images of cardiovascular elements with freeze etching. *Proc. Electron Microscopy Soc. Am.* p. 118 (1967).
64. Leduc, E. H., Bernhard, W., Holt, S. J., and Tranzer, J. P. Ultrathin frozen sections. II. Demonstration of enzyme activity. *J. Cell Biol.* **34**, 773 (1967).
65. Lonsdale, K. The structure of ice. *Proc. Roy. Soc.* **A247**, 424 (1958).
66. Lovelock, J. E. The mechanism of the protective action of glycerol against haemolysis by freezing and thawing. *Biochim. Biophys. Acta* **11**, 28 (1953).
67. Lovelock, J. E., and Bishop, M. W. H. Prevention of freezing damage to living cells by dimethyl sulphoxide. *Nature* **183**, 1394 (1959).

68. Luft, J. H. Permanganate. A new fixative for electron microscopy. *J. Biophys. Biochem. Cytol.* **2**, 799 (1956).
69. MacKenzie, A. P., and Rasmussen, D. H. Low temperature studies with reference to conditions commonly used in "freeze-etching." *Biophys. J.* **9**, A193 (1969).
70. Malhotra, S. K. A study of the structure of the mitochondrial membrane system. *J. Ultrastruct. Res.* **15**, 14 (1966).
71. Malhotra, S. K. Freeze-substitution and freeze-drying in electron microscopy. *In* "Cell Structure and its Interpretation" (S. M. McGee-Russell and K. F. A. Ross, eds.), p. 11. Arnold, London, 1968.
72. Malhotra, S. K., and Van Harreveld, A. Some structural features of mitochondria in tissues prepared by freeze-substitution. *J. Ultrastruct. Res.* **12**, 473 (1965).
73. Malmstrom, B. G. Theoretical considerations of the rate of dehydration by histological freeze drying. *Exptl. Cell Res.* **2**, 688 (1951).
74. Mazur, P. Causes of injury in frozen and thawed cells. *Federation Proc.* **24**, Suppl. 15, S-175 (1965).
75. McAlear, J. H., and Kreutziger, G. O. Freeze etching with radiant energy in a simple cold block device. *Proc. Electron Microscopy Soc. Am.* p. 116 (1967).
76. McNutt, N. S., and Weinstein, R. S. Interlocking subunit arrays forming nexus membranes. *Proc. Electron Microscopy Soc. Am.* p. 330 (1969).
77. Meryman, H. T. Replication of frozen liquids by vacuum evaporation. *J. Appl. Phys.* **21**, 68 (1950) (abstr.).
78. Meryman, H. T. Physical limitations of the rapid freezing methods. *Proc. Roy. Soc.* **B147**, 452 (1957).
79. Meryman, H. T. Freezing and drying of biological materials. *Ann. N.Y. Acad. Sci.* **85**, 501, 1960.
80. Meryman, H. T., ed. "Cryobiology." Academic Press, New York, 1966.
81. Meryman, H. T., Freeze-drying. *In* "Cryobiology" (H. T. Meryman, ed.), p. 610. Academic Press, New York, 1966.
82. Meryman, H. T., and Kafig, E. The study of frozen specimens, ice crystals and ice crystal growth by electron microscopy. *Res. Rept. Naval Med. Res. Inst.* (*Bethesda, Md.*) **13**, 529 (1955).
83. Monroe, H. G., Gamble, W. J., La Farge, C. G., Gamboa, R., Morgan, C. L., Rosenthal, A., and Bullivant, S. Myocardial ultrastructure in systole and diastole using ballistic cryofixation. *J. Ultrastruct. Res.* **22**, 22 (1968).
84. Moor, H. Die Gefrier-Fixation Lebender Zellen und ihre Anwedung in der Elektronenmikroskopie. *Z. Zellforsch. Mikroskop. Anat.* **62**, 546 (1964).
85. Moor, H. Ultrastructuren in Zellkern der Bakerhefe. *J. Cell Biol.* **29**, 153 (1966).
86. Moor, H. Use of freeze-etching in the study of biological ultrastructure. *Intern. Rev. Exptl. Pathol.* **5**, 179 (1966).
87. Moor, H., Mühlethaler, K., Waldner, M., and Frey-Wyssling, A. A new freezing ultramicrotome. *J. Biophys. Biochem. Cytol.* **10**, 1 (1961).

88. Moor, H., and Mühlethaler, K. Fine structure in frozen-etched yeast cells. *J. Cell Biol.* **17**, 609 (1963).
89. Moor, H., Ruska, C., and Ruska, H. Elektronenmikroskopische Darstellung tierischer Zellen mit der Gefrierätztechnik. *Z. Zellforsch. Mikroskop. Anat.* **62**, 581 (1964).
90. Mühlethaler, K., Moor, H., and Sjarkowski, J. The ultrastructure of the chloroplast lamellae. *Planta* **67**, 305 (1966).
91. Müller, H. R. Gefriertrocknung als Fixierungsmethode an Pflanzenzellen. *J. Ultrastruct. Res.* **1**, 109 (1957).
92. Mundkur, B. Electron microscopical studies of frozen dried yeast. *Exptl. Cell Res.* **34**, 155 (1964).
93. Parkes, A. S. Discussion meeting. *Proc. Roy. Soc.* **B147**, 423 (1957).
94. Parkes, A. S., and Smith, A. U., eds., "Recent Research in Freezing and Drying." Blackwell, Oxford, 1960.
95. Parsons, D. F. Negative staining of thinly spread cells and associated virus. *J. Cell Biol.* **16**, 620 (1963).
96. Pease, D. C. Eutectic ethylene glycol and pure propylene glycol as substituting media for the dehydration of frozen tissue. *J. Ultrastruct. Res.* **21**, 75 (1967).
97. Pease, D. C. The preservation of tissue fine structure during rapid freezing. *J. Ultrastruct. Res.* **21**, 98 (1967).
98. Pryde, J. A., and Jones, G. O. The properties of vitreous water. *Nature* **170**, 685 (1952).
99. Rapatz, G., and Luyet, B. Electron microscope study of erythrocytes in rapidly frozen frog's blood. *Biodynamica* **8**, 295 (1961).
100. Rayns, D. G., Simpson, F. O., and Bertaud, W. S. Transverse tubule apertures in mammalian myocardial cells: Surface array. *Science* **156**, 656 (1967).
101. Rebhun, L. I. Applications of freeze-substitution to electron microscope studies of invertebrate oocytes. *J. Biophys. Biochem. Cytol.* **9**, 785 (1961).
102. Rebhun, L. I. Freeze-substitution: Fine structure as a function of water concentration in cells. *Federation Proc.* **24**, Suppl. 15, S-217 (1965).
103. Rebhun, L. I., and Gagne, H. T. Some aspects of freeze-substitution in electron microscopy. *Proc. 5th Intern. Conf. Electron Microscopy, Philadelphia, 1962* Vol. 2, Art L-1 (abstr.). Academic Press, New York, 1962.
104. Remsen, C. C. The fine structure of frozen etched *Bacillus cereus* spores. *Arch. Mikrobiol.* **54**, 266 (1966).
105. Remsen, C. C., and Lundgren, D. G. Electron microscopy of the cell envelope of *Ferrobacillus ferroxidans;* prepared by freezing etching and chemical fixation techniques. *J. Bacteriol.* **92**, 1765 (1966).
106. Remsen, C. C., Valois, F. W., and Watson, S. W. Fine structure of the cytomembranes of *Nitrosocystis oceanus*. *J. Bacteriol.* **94**, 422 (1967).
107. Revel, J-P., and Karnovsky, M. J. Hexagonal array of subunits in intercellular junctions of the mouse heart and liver. *J. Cell Biol.* **33**, C7 (1967).
108. Rey, L. R. Study of the freezing and drying of tissues at very low tempera-

ture. *In* "Recent Research in Freezing and Drying" (A. S. Parkes and A. U. Smith, eds.), p. 40. Blackwell, Oxford, 1960.

109. Seno, S., and Yoshizawa, K. Electron microscope observations on frozen dried cells. *J. Biophys. Biochem. Cytol.* **8,** 617 (1960).

110. Simpson, W. L. An experimental analysis of the Altmann technique of freeze-drying. *Anat. Record* **80,** 173 (1941).

111. Sjöstrand, F. S. Electron microscopic examination of tissues. *Nature* **151,** 725 (1943).

112. Sjöstrand, F. S. Systems of double membranes in the cytoplasm of certain tissue cells. *Nature* **171,** 31 (1953).

113. Sjöstrand, F. S. A new ultrastructural element of the membranes in mitochondria and of some cytoplasmic membranes. *J. Ultrastruct. Res.* **9,** 340 (1963).

114. Sjöstrand, F. S., and Baker, R. F. Fixation by freeze-drying for electron microscopy of tissue cells. *J. Ultrastruct. Res.* **1,** 239 (1958).

115. Sjöstrand, F. S., and Elfvin, L-G. The granular structure of mitochondrial membranes and of cytomembranes as demonstrated in frozen-dried tissue. *J. Ultrastruct. Res.* **10,** 263 (1964).

116. Smith, A. U. "Biological Effects of Freezing and Supercooling." Williams & Wilkins, Baltimore, Maryland, 1961.

117. Someda, K., and Weinstein, R. S. The distribution of a particulate component of membranes in brain and retina. *Proc. Electron Microscopy Soc. Am.* p. 200 (1967).

118. Staehelin, A. Die Utrastruktur der Zellwand und des Chloroplasten von *Chlorella. Z. Zellforsch Mikrosop. Anat.* **74,** 325 (1966).

119. Staehelin, L. A. Chloroplast fibrils linking photosynthetic lamellae. *Nature* **214,** 1158 (1967).

120. Staehelin, L. A. The interpretation of freeze-etched artificial and biological membranes. *J. Ultrastruct. Res.* **22,** 326 (1968).

121. Staehelin, L. A. Ultrastructural changes of the plasmalemma and the cell wall during the life cycle of *Cyanidium caldarium. Proc. Roy. Soc.* **B171,** 249 (1968).

122. Staehelin, L. A., Mukherjee, T. M., and Wynn Williams, A. Freeze-etch appearance of the tight junctions in the epithelium of small and large intestine of mice. *Protoplasma* **67,** 165 (1969).

123. Steere, R. L. Electron microscopy of structural detail in frozen biological specimens. *J. Biophys. Biochem. Cytol.* **3,** 45 (1957).

124. Steere, R. L. Freeze-etching simplified. *Cryobiology* **5,** 306 (1969).

125. Stephenson, J. L. Caution in the use of liquid propane for freezing biological specimens. *Nature* **174,** 235 (1954).

126. Stephenson, J. L. Ice crystal growth during the rapid freezing of tissues. *J. Biophys. Biochem. Cytol.* **2,** Suppl. 45 (1956).

127. Stephenson, J. L. Fundamental physical problems in the freezing and drying of biological materials. *In* "Recent Research in Freezing and Drying" (A. S. Parkes and A. U. Smith, eds.), p. 121. Blackwell, Oxford, 1960.

128. Stirling, C. E., and Kinter, W. B. High resolution radioautography of galactose-^{3}H accumulation in rings of hamster intestine. *J. Cell Biol.* **35**, 585 (1967).
129. Stowell, R. E. Cryobiology. *Federation Proc.* **24**, Suppl. 15 (1965).
130. Van Harreveld, A., and Crowell, J. Electron microscopy after rapid freezing on a metal surface and substitution fixation. *Anat. Record* **149**, 381 (1964).
131. Van Harreveld, A., Crowell, J., and Malhotra, S. K. A study of extracellular space in central nervous tissue by freeze substitution. *J. Cell Biol.* **25**, 117 (1965).
131a. Weinstein, R. S. Personal communication (1967).
132. Weinstein, R. S., Abbis, T. P., and Bullivant, S. The use of double and triple uranyl salts as electron stains. *J. Cell Biol.* **19**, 74a (1963).
133. Weinstein, R. S., and Bullivant, S. The application of freeze cleaving techniques to studies of red blood cell fine structure. *Blood* **29**, 780 (1967).
134. Weinstein, R. S., and Merk, F. B. Periodicity in the cytoplasm of freeze-cleaved sheep erythrocytes. *Proc. Soc. Exptl. Biol. Med.* **125**, 38 (1967).
134a. Weinstein, R. S., and Merk, F. B. Unpublished data (1967).
135. Weinstein, R. S., and Someda, K. Freeze-etching of fracture faces of frozen packed red cells with a modified Bullivant-Ames freeze-fracture and replication apparatus. *J. Cell Biol.* **35**, 190a (1967).
136. Weinstein, R. S., and Someda, K. Artifacts of freeze-cleave (freeze-etch) techniques. 1. Pseudomembranes. *Anat. Record* **160**, 448 (1968) (abstr.).
137. Weinstein, R. S., and Someda, K. The freeze cleave approach to the ultrastructure of frozen tissues. *Cryobiology* **4**, 116 (1967).
138. Weinstein, R. S., and Williams, R. G. Freeze cleaving of red cell membranes in paroxysmal nocturnal hemoglobinurea. *Blood* **30**, 785 (1967).
138a. Weinstein, R. S., and Williams, R. G. Unpublished data (1967).
139. Weinstein, R. S., Ivanetich, L., and Nash, G. Replicas of freeze-cleaved tubular myelin figures. *Proc. Electron Microscopy Soc. Am.* p. 134 (1968).
140. Williams, R. C. The application of freeze drying to electron microscopy. *In* "Biological Applications of Freezing and Drying" (R. J. C. Harris, ed.), p. 303. Academic Press, New York, 1954.

Chapter IV

REDUCING THE EFFECT OF SUBSTRATE NOISE
IN ELECTRON IMAGES
OF BIOLOGICAL OBJECTS

W. W. HARRIS*

I. Introduction

A continuing objective of electron microscopy is to resolve successively smaller objects, approaching, in the limiting instance, single
atoms in amorphous structures. The knowledge to be gained by the direct visualization of biological and other macromolecules at the atomic
level is enormous, and considerable thought and effort have been ex-

* Research supported by the Natitonal Cancer Institute, the National Institute
of General Medical Sciences, the National Institute of Allergy and Infectious
Diseases, and the US Atomic Energy Commission.

147

pended to determine whether this objective can indeed be reached. The question of whether instruments can ever be built which would allow the base sequence of DNA or the amino acid sequence of a peptide to be observed directly is not considered here. Rather, we are concerned with the problem of whether or not atom resolving microscopes would be useful in practice because of the difficulties inherent in specimen preparation. It is evident that samples must somehow be supported in space, and unless either the sample is the support or is on the edge of a support (and hence free in space), both the sample and the support or substrate will contribute to the image.

The ideal substrates for electron microscopy would have low mass absorption of electron, no resolvable structure, be easy to produce, and stable under electron bombardment. The first supports used in practice were cellulose–nitrate or cellulose–acetate films such as had been used to support electron diffraction specimens (17) and which are still used for that purpose. Unfortunately, these collodion films, although only 100–200 Å thick and with structures below the resolution limit of early microscopes, are unstable in an electron beam. In 1942, Schaefer and Harker (23) introduced polyvinyl formal replicas. These Formvar* films are thicker than cellulose–nitrate or cellulose–acetate films but withstand electron bombardment to a much greater degree and make good substrates. These were the substrates which were most generally used until the introduction of carbon films (1) and, later, carbon-coated Formvar.

Plastic films such as the foregoing all show structure estimated (3) to be about 100 Å. Observable structure in thin (10–20 Å) carbon films is much smaller; according to Fernández-Morán (5), it is near the limit of resolution of the present electron microscope.

Resolving power of present day electron microscopes is within a factor of two or three of the residue repeat distance in an α-helix (1.5 Å). Anticontamination devices and improved vacuum systems make it possible to observe specimens nearly free of carbonaceous contaminating films formed by interaction of electrons with residual vapors. This results in improved contrast, and, neglecting spherical and chromatic aberration, leaves the object–substrate combination as the major factor limiting resolution.

* Formvar is a registered trademark.

II. Effect of Substrate on Image

The effect of an ideally structureless substrate would be to attenuate the electron beam exponentially in proportion to film thickness. The electrons in a 50 kV beam incident on a carbon film of 10 μg/cm^2, after passing through the film, have the following composition (*28*).

(1) Thirty-four percent are transmitted without deviation and focused at the rear of the objective lens as a demagnified image of the source.

(2) Fifty-five percent interact with the specimen and, thus, lose energy. Of these inelastically scattered electrons, about 80% fall within a cone with a semiaperture angle of 3×10^{-4} radians (*19*). These electrons have lost about 20 eV for the case considered here.

(3) The remaining 11% are scattered without significant energy loss, and are not normally included in the image.

The first group of electrons forms part of the background in the plane of the image. The inelastically scattered electrons are those that have interacted with electrons of the carbon atoms in ionization processes; the elastically scattered electrons are those that have been deflected by the positive nuclei of the carbon atoms.

The mass thickness at which each electron of a given energy undergoes on the average a single scattering event has been termed the clearing thickness (*29*). Such a specimen will transmit 36.8% $(1/e)$ of the primary beam with no interaction, and these electrons produce the bright background in the image. Some of the inelastically and the elastically scattered electrons make up the image.

The ratio of inelastic electrons n_i to elastically scattered electrons n_e varies with atomic number, z, approximately as (*32*)

$$n_i = \frac{25}{z} n_e \tag{1}$$

For $z = 25$, $n_i = n_e$; and for $z < 25$, $n_i > n_e$; thus, for objects of biological interest, $n_i \simeq$ 3–4 times n_e.

The number of interactions which an electron may undergo is unity or larger for a single row of atoms and increases with increasing atom layers, therefore, after 20 or so rows have been traversed, a fraction of incident electrons will have been scattered many times.

These severally scattered electrons appear to come from points in the object plane differing from the initial scattering point, and there will be a decrease in contrast unless a small objective aperture is used. The number of inelastically scattered electrons increases with thickness; these heteroenergetic electrons also reduce image sharpness due to chromatic aberration of the objective. The energy spread causes electrons from any one point to be focused over a disc of diameter d_c *(33)*:

$$\frac{d_c}{M} = kf\frac{\Delta V}{V}\,\alpha_o \tag{2}$$

where M is magnification in the image plane, k is a constant slightly less than one, α_o is the effective aperture of the objective, f is the focal length of the objective, and $\Delta V/V$ is the mean energy loss (in eV). Attainable resolution of an electron microscope is limited, except for very thin specimens, by these inelastically scattered electrons. For example, for a 100-nm carbon film and 100-kV beam, chromatic aberration limits resolution to about 3 nm for a typical microscope, a limit larger than that determined by spherical aberration.

III. Effect of Beam Potential

With increased beam potential, chromatic aberration decreases nearly linearly from 0.1 MeV to 1 MeV, so that for a fixed value of resolution, specimen thickness can be increased as the voltage so long as visibility is maintained. The maximum thickness can be defined as that which passes enough electrons through it and the objective aperture to give a useable image. For commonly called amorphous objects, with an expected attenuation of $1/e^2$ or $1/e$ *(4)*, the theoretical thickness limit should be proportional, for a small aperture, to $(V/c)^2$ as the beam potential V is raised (c being the velocity of light). For a given reduction in transmission, object thickness will be three times as thick as 1 MeV as at 0.1 MeV.

The resulting reduction of scattering background in the image plane obtained by operating at 0.2 MeV to 0.5 MeV and using a substrate thickness acceptable for conventional 0.1 MeV microscopy would, in principle, offer significant reduction in background noise, provided that object fine structure of interest exceeded resolvable substrate structure. At these higher beam potentials, phase contrast would become dominant, necessitating new procedures for interpreting images.

IV. Substrate Structure

Substrate noise of a more serious character appears when local discontinuities in the substrate are imaged along with fine structure of an object, i.e., when substrate structure modulates object fine structure.

At magnifications used to micrograph shadow-cast objects, structure arising from roughness of the surface used to cast the substrate film can, after shadowing, obscure wanted detail in the image. Hall's mica substrate and pseudoreplicas (9) produces exceptionally smooth films and in effect eliminates all substrate noise by submerging it in the graininess of a metal coat which limits useful instrumental magnification to about $15,000\times$. This technique has been especially effective in revealing the presence of globular proteins down to molecular weights as low as 13,000—the periodic structure of collagen, DNA strands, polysomes, and fibrinogen to name but a few significant structures.

Metal grain size of shadow-cast objects varies with the metal, evaporation rate, pressure at time of evaporation, and cleanliness of the system. Grain size of evaporated platinum, for example, on a carbon film has been reported to be about 30 Å (11). Such grain prevents full exploitation of today's instrumental resolving powers, which are in the vicinity of 2–3 Å.

Direct examination of any organic substrate at high enough magnification always shows fine structure which varies with focus. Little or no fine structure is noticed in micrographs made at focus at beam potential up to 100 kV; but away from focus, micrographs show random distributions of bright and dark spots (10)—bright if underfocus and dark if overfocus. Ruska (22) studied carbon films of different thicknesses and carbon films overlaid with platinum–iridium with careful through-focus series in increments of 170 Å (Fig. 1). Photographic factors were canceled by recording each focal series on the same plate. For a through-focus series of a 50-Å evaporated carbon

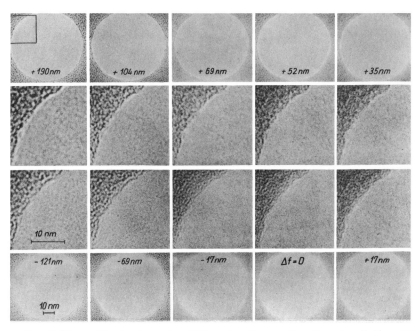

FIG. 1. Film with circular holes (140 nm) with covering film of collodion (10 nm thick) and lightly evaporated platinum–iridium. Through-focus series with specimen cooling to −190°C. 157,000×. Astigmatism = 0. [From Ruska (22).]

film prepared by the method of Bradley (1) and with an objective having little astigmatism, no structure was visible at focus at 80,000×. Ruska concluded that this was because of the absence of astigmatic defocusing and, consequently, absence of phase contrast. Images of a 140-Å carbon film covered with a 100-Å collodion film at 157,000× with 80 kV electrons and negligible astigmatism showed, at focus, disappearance of the weak phase contrast of the thinner covering layer. The phase contrast of the thicker carbon film, although reaching minimum, did not disappear. Groups of platinum–iridium particles from 10–20 Å in size, evaporated over a thin 100-Å collodion layer, could be seen within holes in the 140-Å carbon film but with poor contrast. With optimum focusing on the collodion film, particle contrast was good because phase contrast from the film was at a minimum. Contrast increased with slight underfocusing.

The focus-dependent granulation seen at high magnification is produced by the superposition of the primary wave with waves diffracted by local variations in refractive index for electrons. Lenz and Scheffels (*20*) described the resultant distribution of current density immediately behind the object with the relation

$$a = +(2\lambda\Delta z)^{\frac{1}{2}} \tag{3}$$

between the periodicity a in the object and the defocus Δz which produces maximum contrast for an electron wavelength λ. Thon (*26*) showed that Eq. (3) does not correctly describe the defocusing dependence actually observed when the image structures of high contrast correspond to distances less than 12 Å. However, by taking spherical aberration into account, he was able to explain experimental results for spacing smaller than 12 Å. Experimentally, image structures with distances of 4–5 Å were observed on carbon films over the entire defocus range of $+1000$ to $+5000$ Å, as well as spacings up to 11 Å at a defocus of $+1100$ Å. Spacings less than 10 Å were eliminated from the image when the objective aperture was reduced to 10 μ. Thus, for objects with coarse details producing sufficient scattering absorption contrast, the phase structure of the substrate can be almost completely suppressed, but with a corresponding loss of resolution in the object.

Thon's observations are particularly significant when we consider imaging biological molecules. The 4- to 5-Å substrate image structures, which persisted from -1000 Å to $+1000$ Å defocus and 11-Å spacings at $+1100$ Å, encompass the entire range of significant fine structures of nucleic acids and proteins. These same spacings occur in defocused images of self-supporting protein films. In-focus images show little contrast, but granulation is still present due to defocusing effects caused by lens aberrations.

V. Self-Supporting Films

Jaffe (*16*) partially eliminated the substrate by using a plastic net produced by condensing water vapor on a wet Formvar film. Later Sjöstrand (*25*) described a technique for preparing Formvar nets with fairly uniform holes of proper size for supporting thin sections. Fernández-Morán and Finean (7) coated copper, titanium, and platinum grids with fenestrated Formvar films with uniformly distributed holes of 100–4000 Å diameter. They were prepared by coating either clean glass slides or freshly cleaved mica with a 0.1 or 0.2% Formvar in 1,2-dichloroethylene solution in an atmosphere of controlled humidity. Films were stabilized with a carbon layer by vacuum evaporation. Ten- to 50-Å thick carbon films, evaporated onto cleaved mica in an ion-pumped evaporator at pressures of 10^{-8} to 10^{-9}, were floated onto clean distilled water and picked up on the carbon-coated fenestrated films. The authors state carbon films as thin as 10 Å can be prepared which are nearly structureless.

Proteins and some synthetic polymers form self-supporting films thin enough for electron microscopy but sometimes have a tendency to charge in the electron beam. Heidenreich (14) formed poly-L-glutamate films by first spreading a chloroform solution of the polymer onto a microscope slide and blowing the films onto a bare grid with high pressure air. Incorporation of graphite into the glutamate solution sufficed to reduce sample charging in the electron beam. Harris and Ball (12) prepared stable films of a variety of proteins and amino acids in the following way: A small drop of an aqueous solution of a purified protein fraction is placed on a bare 400-mesh copper grid clamped in tweezers. The drop diameter should be less than the diameter of the grid. The grid and a microscope specimen holder are placed in a deep freeze at −65°F. After 10 minutes the grid is inserted in the specimen holder and placed in the microscope, where it is immediately pumped. Film thickness can be varied by varying the concentration of the protein solution. Although pure amino acids do not easily form thin films by this method, one can usually find thin films protruding from a grid wire which give clear images with 80 kV electrons.

Phase images of unsupported myoglobin films differ distinctly from those of amino acids. The defocused image of glycine, for example, shows no systematic arrangement of electron scattering centers. The images at 11,000,000× consist of irregular bright spots varying from 2 to 7 Å in diameter. Some are isolated but the majority appear to be in continuous contact forming short rods or chains up to 20 Å in length. The separation between spots and chains averages 7 Å.

Myoglobin images differ from the glycine image both dimensionally

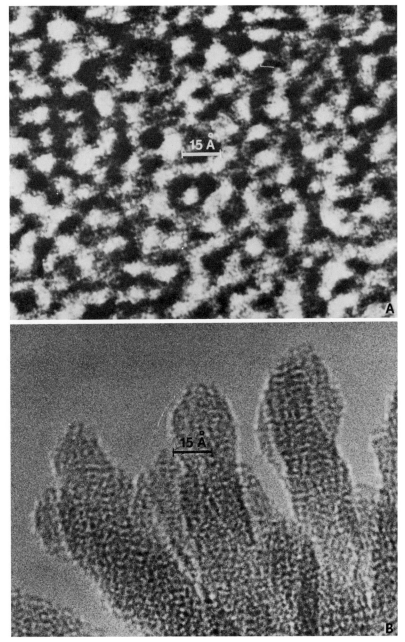

Fig. 2. (A) Myoglobin and (B) chymotrypsin; both are unstained and unsupported. 11,000,000×.

and in apparent ordering. As may be seen in Fig. 2A, the defocused
myoglobin image is composed of white centers from 4–10 Å in
diameter, each of which is made up of bright spots measuring from
1.8 to 2.3 % in diameter. The 6- to 10-Å spots are, on the average,
in continuous contact making tortuous paths within the image. At a
defocus of 210 Å, these columns frequently form nearly closed loops.
A similar type of image is shown in Fig. 2B.

Occasionally small particles such as ferritin can, as shown by
Fernández-Morán (6) in Fig. 3, be attached to asbestos fibers and,
thus, examined without an interfering substrate.

Fig. 3. Unstained ferritin molecule attached to asbestos fiber. [From Fernández-
Morán (5).]

VI. Ideal Substrate

Although thin carbon substrates can be nearly structureless when
viewed with contemporary microscopes, they will not be so when
viewed at the 1-Å level or better. For specimens that cannot be made
self-supporting, an alternative may be to use an ordered substrate
such as a layered graphite crystal.

Since the image of the substrate is superimposed on that of the object, the wanted information has to be extracted from the unwanted. The elastically scattered electrons within the objective aperture lens and transmitted electrons, upon which phase contrast depends, are attenuated by both specimen and object; contrast of the molecular image depends critically upon thickness and scattering power of the substrate. Eisenhandler and Siegel (*4*) analyzed the effect of substrate on the image and concluded that the substrate should be well-ordered and regular, only a few atoms thick, and composed of atoms of low scattering cross-sections and having low atomic numbers.

Image figures for graphite substrates ranging in thickness from one to nine atomic layers were computed with and without attenuation of incident and scattered waves, and the two results compared. If attenuation by the substrate and that by the object do not differ, and if the object is 10%, the image figures for the object will be as though unsupported. If there is a difference in attenuation, object contrast at the points of difference will be reduced. Results were that a layered graphite substrate 10 Å thick or less would not affect contrast of single atoms of C, Na, S, Ni, or Au as computed in the absence of substrate(s), assuming 170 kV accelerating voltage.

Realization of this will require a perfect graphite crystal not more than 10 Å thick with no thickness variation within the field of view. It must be supported in a way such that it is unbent and stable to electron irradiation. Scherzer (*24*) cites examples of images produced by distorted crystals or crystals containing microscopic discontinuities.

VII. Optical Means for Removing Noise

Optical means have been employed to suppress random noise in images of shadowed or stained preparations. A random signal and a superimposed periodic signal can be manipulated photographically so as to reinforce the periodic signal by translation and superimposition of a series of photographs. For objects with radial symmetry such as tobacco mosaic virus discs, Markham (*21*) used a rotating turntable. In use it is rotated at a constant speed of a few hundred rpm, depending upon limitations of ancillary equipment. An enlarged micrograph of the object is placed on the turntable and illuminated with a variable frequency strobe light. Flash frequency is adjusted until symmetry is observed and photographs are made. If n is the number of symmetry elements in the image, n rotations of $360°/n$ will cause all radial symmetry elements to superimpose and reinforce. Random noise effects could be reduced by $(n)^{1/2}$.

Linear periodicities can also be enhanced by the superimposition, either by translating and photographing an enlarged micrograph mechanically or by making 20 or more micrographs of the same periodic object (27) at the same focal setting and then superimposing the negatives in an enlarger and making a composite print.

More recently findings from information theory have been applied to reading electron micrographs. The method as discussed by Fiskin and Beer (8) is based upon work of W. Meyer-Eppler and G. Dorius and makes use of autocorrelation functions of photographs. Physically, an autocorrelation function is a measure of the similarity of two patterns obtained one from the other by a linear displacement and is the average value of the product of the transmissions of the two patterns. The product of the two patterns will peak at displacements corresponding to periodicities in the original pattern. If the transmission function is the sum of periodic and nonperiodic patterns, such as regular object on a random background, the autocorrelation function will show the periodic pattern but with reduced contrast.

Fiskin and Beer (8) obtained autocorrelation of micrographs of shadowed stained polyuridylic acid on carbon–Formvar substrates in the following way: The negative to be analyzed (A) and one that is identical but at reduced magnification (B) are placed parallel to one another at distance a but with centers aligned. An unexposed plate C is placed distance b from the reduced copy is reduced in the ratio $b/a + b$. The system is illuminated with a diffuse source from A through B to C. The intensity at C is directly proportional to the two-dimensional autocorrelation function give by Meyer-Eppler for the function $T\,(x,y)$

$$\varphi(\zeta,n) = \int_{-\infty}^{+\infty} T(x,y)T(x - \zeta, y - n)\, dx\, dy \qquad (4)$$

where the displacement is $\zeta_i + n_i$. Fiskin and Beer (8) were able to detect periodicities of 7–8 Å in shadowed preparations of polyuridylic acid.

The preceding image analysis method depended on the presence of a heavy metal marker at a specific molecular site and relied upon known or anticipated symmetry as do the superposition methods of Markham (*21*) and Valentine (*27*). Substrate noise in micrographs can be minimized electronically as shown by Hart (*13*) with the aid of a flying-spot scanner and a digital computer. Foreknowledge of object symmetry is not needed to process the images. In Hart's method, a series of micrographs, taken at a fixed specimen tilt of 20°, was made with small azimuthal rotations of the specimen about the axis of the microscope between exposures. An identical field of each of these pictures was printed on a single lantern slide to form an array of images each of which represented a different azimuthal aspect of the specimen. With the aid of a flying-spot scanner and a digital computer, a montage was prepared from this plate. For images of tobacco mosaic virus and colloidal gold, the composite picture consisted of 12 images in a 3×4 array covering a 6-cm square. After being corrected for parallax, the images were read with flying-spot scanner which was arranged to read optical density in terms of the difference in intensities of output of two photomultipliers which was fed into a digital computer. The final montage was printed out on a photographic plate by the computer-directed flying-spot scanner. The flying-spot scanner measured the transmitted light intensity and compared it with the incident light intensity. The difference between the logarithms of the intensities is a measure of the optical density of the composite plate at the point being scanned.

Output of the scanner was fed into a digital computer which determined the coordinates of a spot on the scanner cathode ray tube, produced a pulse of light at that spot, and finally received a digitized signal proportional to the difference in intensity of incident and transmitted light. Following steps to relate coordinates of the scanning system to those of the composite of micrographs, the composite is scanned to cover an array of 1023×767 specimen points. Optical density signals numbering 10^7 were recorded on magnetic tape as numbers arranged sequentially according to specimen position and frame number.

Reduction of the data included conversion of measured optical densities of the scan points to a gray value on a scale ranging from 0 to 7 by using the computer. From a transcription tape of the gray values, the computer equalized the twelve images of the composite with respect to overall density and contrast so that they could be superimposed, either optically or by computer. Hart calls the resulting picture a "polytropic montage," and with it he has obtained high-contrast, noise-free images of unstained, unshadowed tobacco mosaic virus.

VIII. Conclusion

The limited number of pertinent references herein cited attests to the fact that the problem of substrate noise has been of limited tractability, although encountered by most microscopists at one time or another. The mica substrate technique (9), the freeze-drying technique of Westerberg (31), and the spreading method of Kleinschmidt et al. (18) for DNA all utilize metallizing to develop contrast and to cancel substrate yet introduce a different noise. Electron stains, such as uranium, molybdenum, lead, and tungsten, as used with thin sections, dispersed particles, or spread films, all enhance the signal-to-noise ratio but with consequent sacrifice of resolving power.

Fenestrated films afford a compromise as does the technique of adsorbing particulates onto a fiber (5) which offers limited areas of unsupported object. Spread films such as shown in Figs. 1 and 2, although free of substrate, are likely to be of variable thickness and break up easily when thin. Very dilute amino acid solutions, handled with the drop freezing technique, break up into droplets adhering to the grid wires. Useable information in this event is restricted to the very outer edges of the object, which would be but a few mono-layers thick.

Today we can envision resolution improvements by use of careful voltage control at higher beam energies, as well as correction of spherical aberration, which will require corresponding improvements in the substrate if any real gains are to result. Thin graphite films appear to offer the lowest possible noise level and should receive a large development effort (4). Even with low substrate noise, the statistical error enforced by radiation damage to interesting biological objects such as macromolecules probably requires selective staining by reliably placed heavy atoms as a means of enhancing the otherwise low contrast (*2, 15, 30*). With these provisions the extraction of useful DNA sequence information by electron microscopy becomes a real possibility for the future.

REFERENCES

1. Bradley, D. E. Evaporated carbon films for use in electron microscopy. *Brit. J. Appl. Phys.* **5**, 65 (1954).
2. Butler, J. W. Digital-computer applications of electron microscopy. *Proc. AMU (Assoc. Midwest Univ.)—ANL (Argonne Natl. Lab.) Workshop High-Voltage Electron Microscopy, 1960* ANL-7275 Instr. (TID-4500) AEC Res. Develop. Rept., pp. 174–176. 1966.
3. Calbick, C. J. Inorganic replication in electron microscopy. *Bell System Tech. J.* **30**, 803 (1951).
4. Eisenhandler, C., and Siegel, B. Effect of the substrate on image contrast in resolution. *Proc. 25th Anniv. Meeting Electron Microscopy Soc. Am.* p. 232. Claitor's Book Store, Baton Rouge, Louisiana, 1967.
5. Fernández-Morán, H. Electron microscopy in the future. *Proc. 25th Anniv. Meeting Electron Microscopy Soc. Am.* p. 11. Claitor's Book Store, Baton Rouge, Louisiana, 1967.
6. Fernández-Morán, H. High resolution electron microscopy of biological specimens. *Proc. 6th Intern. Conf. Electron Microscopy, Kyoto, Japan, 1966* pp. 13–14. Maruzen Co., Ltd., Tokyo, 1966.
7. Fernández-Morán, H., and Finean, J. P. Electron microscope and low-angle X-ray diffraction studies of the nerve myelin sheath. *J. Biophys. Biochem. Cytol.* **3**, 725 (1957).
8. Fiskin, A. M., and Beer, M. Autocorrelation functions of noisy electron micrographs of stained polynucleotide chains. *Science* **159**, 1111 (1968).
9. Hall, C. E. In studies on biological macromolecules. *In* "Modern Developments in Electron Microscopy" (B. M. Siegel, ed.), pp. 395–415. Academic Press, New York, 1964.
10. Hall, C. E. Image characteristics. *In* "Introduction to Electron Microscopy" (C. E. Hall, ed.), pp. 296–307. McGraw-Hill, New York, 1953.
11. Hamlet, R. G. "The Power Spectra of Evaporated Metal Films," Rept. No. 161, p. 70. Mater. Sci. Center, Cornell University, Ithaca, New York, 1964.
12. Harris, W. W., and Ball, F. High resolution electron microscopy of antigens. *Federation Proc.* **27**, No. 2, 366 (1968).
13. Hart, R. G. Electron microscopy of unstained biological material; the polytropic image. *Science* **15**, 1464 (1968).
14. Heidenreich, R. D. Electron phase contrast images of molecular detail. *Siemens Rev.* **34**, 4 (1967).

15. Highton, P., and Beer, M., The minimum mass detectable by electron microscopy. *J. Roy. Microscop. Soc.* [3] **88**, Part 1, 23 (1968).
16. Jaffe, M. Auxillary supporting nets for fragile electron microscope preparations. *J. Appl. Phys.* **19**, 1191 (1948).
17. Kirchner, F. Interferenzapparat für Demonstration und Strukturuntersuchungen. *Physik. Z.* **31**, 772 (1930).
18. Kleinschmidt, A., Gehatia, M., and Zalm, R. K. Uber die molekular Morphologie von desoxribonuclein Sauren. *Kolloid-Z.* **169**, 156 (1960).
19. Lenz, F. Scattering of medium energy electrons at very small angles. *Z. Naturforsch.* **9a**, 185 (1957).
20. Lenz, F., and Scheffels, W. Das Zusammenwirke von phasen und amplituden Kontrast in der electroner mikroskopischen Abbilding. *Z. Naturforsch.* **13a**, 226 (1958).
21. Markham, P., Frey, S., and Hills, G. J. Methods for enhancement of image detail and accentuation of structure in electron microscopy. *Virology* **20**, 88 (1963).
22. Ruska, E. Past and present attempts to attain the resolution limit of the transmission electron microscope. *In* "Advances in Optical and Electron Microscopy" (V. E. Cosslett and R. Barer, eds.), Vol. I, pp. 115–179 (see p. 171). Academic Press, New York, 1966.
23. Schaefer, V.-J., and Harker, D. Surface replicas for use in electron microscopy. *J. Appl. Phys.* **13**, 427 (1942).
24. Scherzer, O. Image formation as a problem of wave theory. *In* "Quantitative Electron Microscopy" (E. H. Zeitler and G. F. Bahr, eds.), p. 61. Williams & Wilkins, Baltimore, Maryland, 1965.
25. Sjöstrand, F. S. "An improved method to prepare formvar nets for mounting thin sections for electron microscopy. *Proc. Reg. Conf. (Eur.) Electron Microscopy Stockholm, 1956* p. 120. Almqvist & Wiksell, Uppsala, 1957.
26. Thon, F. On the defocussing dependence of phase contrast in electron microscopial images. *Z. Naturforsch.* **21a**, 476 (1966).
27. Valentine, R. C. Fundamental difficulties in obtaining very high resolution of biological specimens. *Proc. 3rd Reg. Conf. (Eur.) Electron Microscopy, Prague 1964* p. 23. Publ. House Czech. Acad. Sci., Prague, 1965.
28. van Dorsten, A. C. Role of acceleration voltage. *In* "Quantitative Electron Microscopy" (E. H. Zeitler and G. F. Bahr, eds.), p. 81. Williams & Wilkins, Baltimore, Maryland, 1965.
29. von Borries, B. Electron scattering and image formation in the electron microscope. *Z. Naturforsch.* **4a**, 51 (1949).
30. Welton, T. *Proc. 27th Anniv. Meeting Electron Microscopy Soc. Am.*, p. 182. St. Paul, Minnesota.
31. Westerberg, E. R. Techniques in the freeze drying of macromolecules for electron microscopy. Master's Thesis, Massachusetts Institute of Technology (1960).
32. Zeitler, E. H., and Bahr, G. F. Contrast and mass thickness. *In* "Quantitative

Electron Microscopy" (E. H. Zeitler and G. F. Bahr, eds.), p. 208. Williams & Wilkins, Baltimore, Maryland, 1965.

33. Zworykin, V. K., Morton, G. A., Ramberg, E. G., Hillier, J., and Vance, A. W. Image formation in the electron microscope. *In* "Electron Optics and the Electron Microscope," p. 720. Wiley, New York, 1948.

Chapter V

AUTOMATION IN TISSUE PROCESSING FOR ELECTRON MICROSCOPY

WILLIAM G. BANFIELD

Those who have spent many hours in processing tissues for electron microscopy are in agreement that a machine is needed for this technically wasteful chore. A machine should be able to do the job better than it is done manually since the changing of fluids can be controlled precisely and tissues changed simultaneously. Agitation, temperature, and even humidity and pressure can be controlled throughout the operation. In addition large numbers of specimens can be processed together. If desired, the processing can be done at night so that the tissues are ready for embedding in the morning or when the schedule permits.

Bernhard (*3*) described a simple continuous drip apparatus for gradually dehydrating a specimen; shakers (*6*) and rotary mixers (*12*) have been used for agitation. However, no automatic tissue processor which allows for precise control of fixation, dehydration, and infiltration of the small specimens used for electron microscopy is readily available.

I. Tissue Processors

Two groups, one in Japan (*1, 2*) and the other in the United States (*9*) have been working independently to develop an automatic tissue processor for electron microscopy. The principles used by each are different. The Japanese developed their machine along the lines of the TECHNICON used for paraffin embedding of light microscopic sections, and the U.S. group, with a new approach, used successive reservoirs emptying into a chamber where the tissues are suspended in a holder.

A. AIHARA PROCESSOR

This instrument is still under development by Aihara, Nishikawa, Osada, Suzuki, and Nakamura, although commercial models are available (*1*). It is divided into two units: an autoprocessor operating on the principle of the TECHNICON and an autocontroller which is synchronized with the autoprocessor which controls the temperature, humidity, and pressure (Figs. 1 and 2). The temperature is regulated by circulating water, the humidity by the cir-

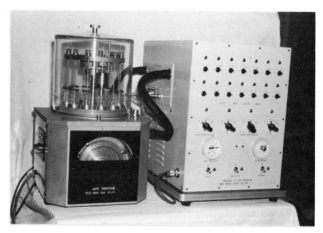

Fig. 1. Aihara processor. Autoprocessor at left, autocontrollor at right.

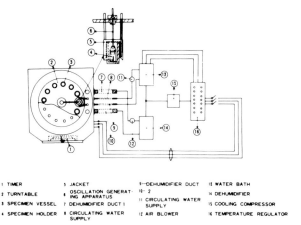

1 TIMER	5 JACKET	9 DEHUMIDIFIER DUCT	13 WATER BATH
2 TURNTABLE	6 OSCILLATION GENERATING APPARATUS	10 2	14 DEHUMIDIFIER
3 SPECIMEN VESSEL	7 DEHUMIDIFIER DUCT I	11 CIRCULATING WATER SUPPLY	15 COOLING COMPRESSOR
4 SPECIMEN HOLDER	8 CIRCULATING WATER SUPPLY	12 AIR BLOWER	16 TEMPERATURE REGULATOR

FIG. 2. Aihara processor. Schematic drawing.

culation of air, and the vacuum by a connection with a vacuum pump. The various conditions can be set on a panel and the tissue processed automatically once the machine is activated. The tissue blocks are carried in a holder several types of which are under consideration. Agitation of the holder is produced by a reciprocating vertical movement of 20–45 cpm through a specimen cage support shaft.

B. NORRIS PROCESSOR

Figures 3–6 illustrate the tissue processor developed by Norris, Banfield, and Chalifoux. [Much of this section was taken verbatem from Norris *et al.* (*9*).]

The tissue is held in small chambers bored in a stainless steel disc. The chambers are covered on top and bottom by a 100-mesh stainless steel screen held in place by steel syringe tubing inserts (Figs. 4 and 5). The disc is split into a top and bottom section. The top section has a spacer (Fig. 4) machined into the upper surface, and on the lower surface the syringe tubing inserts project above the rims of the chambers to make a snug fit into the chambers of the lower section (Fig. 5). This cuff prevents the tissue blocks from floating out of

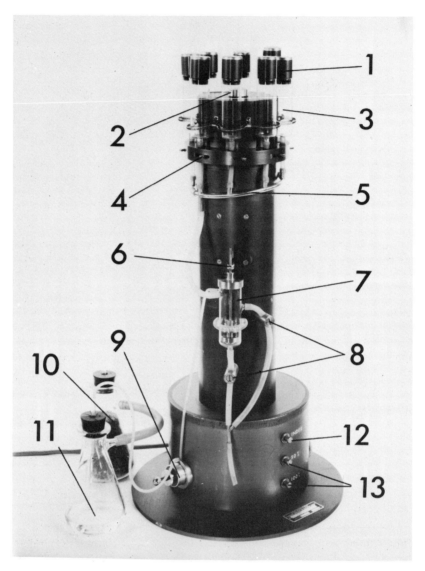

FIG. 3. Processor unit: (1) desiccant caps; (2) station indicator; (3) reservoir; (4) spring loaded cam-actuated tubing clamps; (5) contoured manifold; (6) agitator plug-in; (7) mixing chamber; (8) one revolution motor driven cam-operated clamps; (9) variable controlled peristaltic pump; (10) silica gel; (11) reservoir of dehydration liquid; (12) manual station advance button; and (13) buttons for 50% and complete emptying of mixing chamber.

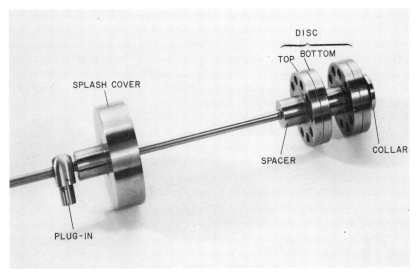

Fɪɢ. 4. Agitator rod with plug-in, splash cover, and assembled tissue holder discs held against the stop pin by the spring loaded collar. 0.77×.

the chambers, forms a convenient well in which to place the blocks on loading the chambers, and holds the blocks when the two sections are disengaged following processing. Top and bottom chambers are numbered, and they are grooved on the side for aligning. The disc is slid onto the lower part of a shaft in which there is a stop pin. The present machine is equipped with two discs, each with 10 chambers. Each chamber will hold five or more tissue blocks. The discs are held snugly against the stop pin by a spring loaded collar. On the shaft above the stop pin is a splash lid, which is prevented from sliding off the shaft at the top by a pin which plugs into the agitating arm of the tissue changer. The shaft with discs is suspended in a glass mixing chamber with three side arms and an outlet at the bottom. Two of the side arms are close to the top and used as inlets, and one is placed in a lower position so that half the fluid volume can be drained through it. The mixing chamber is filled to this arm with the first dilution of dehydrating fluid. One hundred percent fluid is added by a peristaltic pump, the speed of which can be regulated. The time desired to fill the remainder of the chamber is set on the

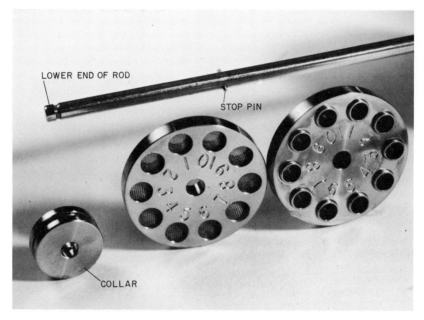

LOWER END OF ROD

STOP PIN

COLLAR

FIG. 5. Tissue holder disassembled. Top section of disc at right. 1.5×.

control panel, and the pump speed regulated correspondingly. When the chamber is full, the side arm opens and drains the chamber half-way. The chamber again fills, and the process can be repeated up to 4 times or terminated after any cycle.

After three cycles the concentration of dehydrating agent is about 95%, if a 65% solution is used at the start. Following the last dilution the chamber drains completely. There are reservoirs at the top of the tissue changer draining into a circular inclined tube (contoured mani-fold) split into two sections to facilitate removal. An outlet at the lowest point of the tube drains into one of the upper side arms of the mixing chamber. The connections are made to the contoured mani-fold with rubber tubing, since the control of fluid is governed by spring loaded hose clamps opened by a motor through a cam arrange-ment. After the last dilution and the emptying of the mixing chamber, the first reservoir opens and drains into the mixing chamber.

The time the tissue stays in the solution can be set on the control panel (Fig. 6). After this time, the mixing chamber automatically empties, then fills from the next reservoir. The time is preset on the control panel for each solution so that the changing is automatic once the dilution cycles begin. The contour manifold is made from stainless steel but could be made from glass. Moisture is removed from air entering the reservoirs by caps filled with a drying compound. Air entering the dehydration reservoir also passes over a drying compound. At the beginning of each dilution cycle, and at each addition of a new solution from a reservoir, the agitation arm is activated and pumps the discs up and down in the mixing chamber solution 100 times per minute for 5 minutes. This is necessary to remove air bubbles trapped in the disc chambers. The machine can be set to leave the mixing chamber full or empty at the termination of the program.

The initial dilution cycles can be bypassed and the program started using the reservoirs. The reservoirs can be emptied manually by pressing a lever beneath each one. The changer can be made to advance manually through a button activating the cam motor. The reservoir position of the cam is indicated by a pointer on the top

Fig. 6. Control panel: dial to the right of reset tab is used for selecting the number of dilution cycles desired. The numbers 2, 3, and 4 indicate the dilution cycle in process. White tabs "O" through "C" are for selecting the reservoir station at which the program is terminated.

of the tissue changer. The changer can be automatically set to the starting position through a reset button on the control panel. The mixing chamber can be emptied manually through a button which will open the lower side arm for 50% emptying or another opening the lower drain for complete emptying.

Within the base and column of the processor are five motors: one for the dilution cycle pump; one for 50% emptying of the mixing chamber; one for complete emptying; one for controlling reservoir emptying; and one for agitation. Dilution cycles are set with a single timer interlocked with reset relays and synchonized with the activating system for the 50% drain motor.

The emptying of the mixing chamber is controlled by a timer, which is automatically synchronized with the emptying of the next reservoir and the motor for agitation. The emptying of a reservoir activates the timer for the next cycle.

The tissue changer has been used in our laboratory to dehydrate tissues and to infiltrate tissues for methacrylate embedding. It can be used for Epon embedding to the 75:25 Epon:propylene oxide mixture. However, the reservoirs containing the Epon should bypass the manifold through a separate rubber tube to an additional side arm of the mixing chamber. The tubing can then be discarded after each use. A pump could be installed for introduction of a final, more viscous solution which would not flow easily by gravity.

Alternatively, heating coils could be placed around the reservoir, tubing and mixing chamber to be activated when needed to lower the viscosity, facilitating the draining of the reservoir, and the penetration of the epoxy embedding medium into the tissue. Although no provisions have been made for vacuum, this also could be applied to the mixing chamber to be activated when needed.

The tissue changer, although designed to process small blocks from 20 specimens, could, in principle, be modified for other sizes of specimen in any number. The tissue block may not be of such a consistency that it is disrupted by the introduction of fluid or the agitation at the beginning of each cycle.

There may be some advantage in the reservoir–mixing chamber machine over the TECHNICON machine if conditions have to be controlled only in the relatively small mixing chamber where the tissues are processed.

II. Fixation, Dehydration, and Embedding

The methods used for preparing the tissue for cutting are many (*5*) and vary depending upon the problem, the tissue, the laboratory, and even individual preference. The variables inherent in the fixation and embedding of tissues and those introduced by cutting, mounting, and electron microscopy resulting in the final pictures used for comparison of methods are so great that it is difficult to compare even two methods without long experience with both simultaneously in the same laboratory and with the same tissues. Even then, what seems to be the best in one laboratory may not prove so in another.

However, it is clear that methacrylate is inferior to the epoxy embedding materials in the preservation of tissue structure. In skin, for instance, methacrylate causes distintegration of collagen bundles and swelling of the elastic fibers in an unpredictable manner. This is not seen with Epon embedding. It is also clear that fixation with glutaraldehyde followed by fixation with osmium tetroxide (*8, 10, 11*) will give better morphology than when glutaraldehyde is used alone and will preserve structures such as microtubules (*7*) and myelin figures in embryonic chick hepatocytes (*4*) lost when osmium is used alone.

III. Future Prospects

Automatic tissue processors for electron microscopy will become commonplace. There will not only be versatile instruments for general use but also those tailored for specifice processes. Some of these could be greatly simplified and made inexpensive in order that there could be several in one laboratory.

REFERENCES

1. Aihara, K., Nishikawa, H., Osada, K., Suzuki, K., and Nakamura, T. Electron microscope automatic tissue processor. *Igaku no Ayumi* **63**, 327 (1967).
2. Aihara, K., Nomizo, K., Nagata, K., Nishikawa, H., and Suzuki, K. An attempt to incorporate the automated tissue processing in electron microscopy. *J. Electronmicroscopy (Tokyo)* **16**, 285 (1967).

3. Bernhard, W. Appareil de deshydratation continue. *Exptl. Cell Res.* **8,** 248 (1955).

4. Curgy, J. Influence du mode de fixation sur la possibilité d'observer des structures myélineques dans les hépatocytes d'embryons de poulet. *J. Microscopie* **7,** 63 (1968).

5. Kay, D., ed., Techniques for Electron Microscopy. Davis, Philadelphia, Pennsylvania, 1965.

6. Kushida, H. Penetration of embedding media into specimens. *J. Electronmicroscopy (Tokyo)* **13,** 107 (1964).

7. Ledbetter, M., and Porter, K. A "Microtubule" in plant cell fine structure. *J. Cell Biol.* **19,** 239 (1963).

8. Millonig, G. *Proc. 5th Intern. Conf. Electron Microscopy, Philadelphia, 1962* Vol. 2, Art P8. Academic Press, New York, 1962.

9. Norris, G., Banfield, W., and Chalifoux, H. Tissue processor for automatic dehydration and infiltration of small specimens. *Sci. Tools* **14,** 13 (1967).

10. Palade, G. A study of fixation for electron microscopy. *J. Exptl. Med.* **95,** 258 (1952).

11. Sabatini, D., Bensch, K., and Barrnett, R. Cytochemistry and electron microscopy. The preservation of cellular ultrastructure and enzymatic activity by aldehyde fixation. *J. Cell Biol.* **17,** 19 (1963).

12. Steinbrecht, R. A., and Ernest, K. Continuous penetration of delicate tissue specimens with embedding resin. *Sci. Tools* **14,** 24 (1967).

Author Index

Numbers in parentheses are reference numbers and indicate that an author's work is referred to although his name is not cited in the text. Numbers in italics show the page on which the complete reference is listed.

A

Abbis, T. P., 114(132), *146*
Adams, C. W. M., 87(1, 2), *93*
Afselius, B. A., 71, 88(3), 91(3), *93*
Agar, A. W., 22(1), 44, *61*
Aihara, K., 166(1, 2), *173*
Albert, L. von, 57, *61*
Ames, A., 120(19), 127, 133(19), 136, *140*
Anderson, O. R., 76(4), 87, *93*
Anderson, P. J., 76(13), *94*
Arnold, E. A., 92(120), *99*

B

Bahr, G. F., 27(3), *61*, 71(11), 72(7, 8, 10, 44), 73, 74(8, 9), 85(6), 86(6), *93, 94, 95*, 149(32), *162*
Baker, J. R., 72(12), *94*
Baker, R. F., 102(114), 110(114), 120, *139, 145*
Ball, F., 154, *161*
Banfield, W., 166(9), 167(9), *174*
Barer, R., 37(4), 51(4), *62*
Barka, T., 76(13), *94*
Barrnett, R. J., 72(14), 91(105), 92(104, 105), *94, 98*, 173(11), *174*
Bayer, M., 87(15), *94*
Becker, H., 7, *62*
Beer, M., 85(16, 50), *94, 95*, 158, *161, 162*
Bennett, A. H., 37, *62*

Bensch, K. G., 91(105), 92(104, 105), *98, 173(11), 174*
Bernhard, W., 70(65), *96, 122, 139, 142,* 165(3), *174*
Bertaud, W. S., 126(4, 100), 137(4), *139, 144*
Bethe, H., *62*
Biberfeld, P., 92(31), *94*
Birbeck, M. S. C., 70(17), *94*
Birch-Anderson, A., 70(18), *94*
Bishop, M. W. H., 106(67), *142*
Bloom, G., 72(8, 10), 73(8, 9, 10), 74(8, 9), *93, 94*
Bloom, M. G., 71(47), *95*
Boersch, H. von, 34, *62*
Bondareff, W., 109, *139, 141*
Bonham, R. A., 43, *62*
Booij, H. L., 77(57), 82, 83, 84, *96, 98*
Born, M., 37(10), 53(10), *62*
Bornemann, F., 71(101), 88(101), *98*
Borries, B. von, 44(11), 50(11), 54, 55 (12), 56(12), *62*, 149(29), *162*
Bradley, D. E., 148(1), 152, *161*
Branton, D., 125, 126(6, 7, 8, 9, 10, 11, 23, 24), 134, 135, 136, 137(8, 23), 138(8, 23), *139, 140*
Bremmer, H., 43(13), 53(13), *62*
Brenner, S., 92(19), *94*
Bullivant, S., 92(20), *94*, 107(15), 108 (15), 112(15), 113(12, 13), 114(13,

G

Gagne, H. T., 113(103), 114, *144*
Gamble, W. J., 117(83), 119(83), *143*
Gamboa, R., 117(83), 119(83), *143*
Gehatia, M., 160(18), *162*
Geren, B. B., 76(37), *95*
Gersh, I., 108, 109, 110(47), *141*
Glaser, W., 4, 8(48, 50), *64*
Glauert, A. M., 91(39), 92(38), *95*
Glick, D., 108, 109, *142*
Glossop, A. B., 34(51), *64*
Gordon, A., 76, *99*
Gottlieb, A., 89(40), *95*
Greenawalt, J. W., 76(49), 88(49), *95*
Griffith, W. P., 71(41), *95*
Grivet, P., 5(52), *64*
Gruber, M., 93(73), *96*
Grunbaum, B. W., 110, *142*

H

Haggis, G. H., 124, *142*
Hagström, L., 72(44), *95*
Haine, M. E., 21, 31, 43(53), *64*
Hair, M. L., 71(42, 43), *95*
Hake, F., 70(45), 86(45), 88(45), 90(45), *95*
Hall, C. E., 34(54, 55), *64*, 92(46), 93(46), *95*, 123, *142*, 151(9, 10), 160(9), *161*
Hallet, J., 104(52), *142*
Hamlet, R. G., 151(11), *161*
Hanker, J. S., 71(47, 109), 72(109), *95*, *98*
Hanszen, K. J., 13(56), *64*
Hanzon, V., 102(54), 110(53, 54), 113, 120, *142*
Harker, D., 148, *162*
Harris, R. J. C., 103(55, 56), *142*
Harris, W. W., 154, *161*
Hart, R. G., 159, *161*
Hart, R. K., 27(57), *64*
Harwood, H. J., 90(48), *95*
Hausner, G., 25(96a), *67*
Hawkes, P. W., 8, *64*

Haydon, G. B., 53(59), *64*
Hecht, F., 89(40), *95*
Heide, H. G., 24, *65*
Heidenreich, R. D., 18(62), 23, 54, 58, 59, *65*, 154, *161*
Heinemann, K., 29, *65*
Hermodsson, L. H., 102(54), 110(53, 54), 113(54), 120, *142*
Heun, F. A., 76(49), 88(49), *95*
Heuser, J. E., 74(28), 91(28), *94*
Hibi, T., 17, 18(66, 120, 121), 35, *65*, *68*
Highton, P. J., 85(50), *95*, *161*, *162*
Hillier, J., 4, *65*, 150(33), *163*
Hills, G. J., 157(21), 159(21), *162*
Hirsch, P. B., 5(68), 34, *65*
Hoelke, C. W., 18(97), *67*
Holt, S. J., 122(64), *142*
Hooghwinkel, G. J. M., 83(51), *95*
Hoppe, W., 59(82), *66*
Horne, R. W., 92(19, 52, 53, 122), *94*, *95*, *99*
Horner, J. A., 22, *66*
Howie, A., 5(68), 34(68), *65*

I

Iro, K., 44(72), *65*
Isenberg, I., 109(45, 46), *141*
Ito, B., 76(91), *97*
Ito, S., 72(54), 76(92), *95*, *97*
Ivanetich, L., 133(139), 136(139), *146*
Iwanaga, M., 27(111), *67*
Iwasa, N., 27(111), *67*

J

Jaffe, M., 153, *162*
Job, P., 77(55), *96*
Johannisson, E., 72(10), 73(10), *94*
Johnson, D. J., 57, *65*
Johnson, H. M., 35, 47(70), 49(91), 50(70), 57, *65*, *66*
Jones, G. O., 104, *144*
Jost, M., 126(57), *142*
Jupnik, H., 37(6), *62*

Subject Index

A

Aberrations of magnetic lenses, 5–16, 27
  AC ripple, 9
  calculation from axial magnetic field
    distribution, 7, 8
  chromatic aberration, 10, 27, 150
  disc of confusion, 7, 10, 15
  dynamic and static, 9
  experimental measurement, 8
  random supply fluctuations, 9
  spherical aberration, 5, 7, 15
  wave aberration, 7
Amorphous ice, 124
Arrays of light detectors, 30
Artifacts, 20, 70, 92, 102
Autocontroller, 166
Autocorrelation functions, 158
Autoprocessor, 166

B

Beam damage, 26
Bremsdicke or clearing thickness, 44,
  149

C

Coherence, 13, 14
Contamination, 24
Contrast, 2–3, 28–31, 37, 49, 57
  definition of, 29–32
  displacement defocus contrast, 57
  electron optical, 2
  eye contrast, 31
  high voltage, 26
  in-focus phase contrast, 28, 37–52
  instrumental contrast, 31
  macrocontrast, 49
  metal contrasting, 2, 28
  metal shadowing, 3
  microcontrast, 49
  negative staining, 2
  out-of-focus phase contrast, 28
  positive staining, 2
  Weber–Fechner law, 31
Coolants, 107–113
  freons, 107
  Genetron, 23, 113
  helium I, 108
  helium II, 108, 112
  isopentane, 108
  liquid nitrogen, 108
  propane, 107
Cooling rate, 107
Cryofixation, 112

D

Damage, 2
  chemical damage, 2
  electron beam, 2
  vacuum drying, 2
Dark field, 32–37
  central back focal plane aperture
    method, 34
  off-center aperture, 34
  Schlieren or Foucault method, 33
  Spierer lens, 34
  strioscopic, 34
  strioscopy, 32